SUBJECTIVE WELL-BEING

Measuring Happiness, Suffering, and
Other Dimensions of Experience

Panel on Measuring Subjective Well-Being in a Policy-Relevant Framework
Arthur A. Stone and Christopher Mackie, *Editors*

Committee on National Statistics

Division of Behavioral and Social Sciences and Education

NATIONAL RESEARCH COUNCIL
OF THE NATIONAL ACADEMIES

THE NATIONAL ACADEMIES PRESS
Washington, D.C.
www.nap.edu

THE NATIONAL ACADEMIES PRESS   500 Fifth Street, NW   Washington, DC 20001

NOTICE: The project that is the subject of this report was approved by the Governing Board of the National Research Council, whose members are drawn from the councils of the National Academy of Sciences, the National Academy of Engineering, and the Institute of Medicine. The members of the panel responsible for the report were chosen for their special competences and with regard for appropriate balance.

This study was supported by Task Order No. N01-OD-42139 between the U.S. National Institutes of Health and the National Academy of Sciences, and award number 10000592 between the UK Economic and Social Research Council and the National Academy of Sciences. Support for the Committee on National Statistics is provided by a consortium of federal agencies through a grant from the National Science Foundation (award number SES-1024012). Any opinions, findings, conclusions, or recommendations expressed in this publication are those of the author(s) and do not necessarily reflect the views of the organizations or agencies that provided support for the project.

International Standard Book Number-13:   978-0-309-29446-1
International Standard Book Number-10:   0-309-29446-0

Additional copies of this report are available from the National Academies Press, 500 Fifth Street, NW, Keck 360, Washington, DC 20001; (800) 624-6242 or (202) 334-3313; http://www.nap.edu.

Copyright 2013 by the National Academy of Sciences. All rights reserved.

Printed in the United States of America

Suggested citation: National Research Council. (2013). *Subjective Well-Being: Measuring Happiness, Suffering, and Other Dimensions of Experience*. Panel on Measuring Subjective Well-Being in a Policy-Relevant Framework. A.A. Stone and C. Mackie, Editors. Committee on National Statistics, Division of Behavioral and Social Sciences and Education. Washington, DC: The National Academies Press.

THE NATIONAL ACADEMIES
Advisers to the Nation on Science, Engineering, and Medicine

The **National Academy of Sciences** is a private, nonprofit, self-perpetuating society of distinguished scholars engaged in scientific and engineering research, dedicated to the furtherance of science and technology and to their use for the general welfare. Upon the authority of the charter granted to it by the Congress in 1863, the Academy has a mandate that requires it to advise the federal government on scientific and technical matters. Dr. Ralph J. Cicerone is president of the National Academy of Sciences.

The **National Academy of Engineering** was established in 1964, under the charter of the National Academy of Sciences, as a parallel organization of outstanding engineers. It is autonomous in its administration and in the selection of its members, sharing with the National Academy of Sciences the responsibility for advising the federal government. The National Academy of Engineering also sponsors engineering programs aimed at meeting national needs, encourages education and research, and recognizes the superior achievements of engineers. Dr. C. D. Mote, Jr., is president of the National Academy of Engineering.

The **Institute of Medicine** was established in 1970 by the National Academy of Sciences to secure the services of eminent members of appropriate professions in the examination of policy matters pertaining to the health of the public. The Institute acts under the responsibility given to the National Academy of Sciences by its congressional charter to be an adviser to the federal government and, upon its own initiative, to identify issues of medical care, research, and education. Dr. Harvey V. Fineberg is president of the Institute of Medicine.

The **National Research Council** was organized by the National Academy of Sciences in 1916 to associate the broad community of science and technology with the Academy's purposes of furthering knowledge and advising the federal government. Functioning in accordance with general policies determined by the Academy, the Council has become the principal operating agency of both the National Academy of Sciences and the National Academy of Engineering in providing services to the government, the public, and the scientific and engineering communities. The Council is administered jointly by both Academies and the Institute of Medicine. Dr. Ralph J. Cicerone and Dr. C. D. Mote, Jr., are chair and vice chair, respectively, of the National Research Council.

www.national-academies.org

PANEL ON MEASURING SUBJECTIVE WELL-BEING IN A POLICY-RELEVANT FRAMEWORK

ARTHUR A. STONE (*Chair*), Department of Psychiatry and Behavioral Sciences, Stony Brook University
NORMAN M. BRADBURN, Department of Psychology, University of Chicago
LAURA L. CARSTENSEN, Department of Psychology, Stanford University
EDWARD F. DIENER, Department of Psychology, University of Illinois at Urbana-Champaign
PAUL H. DOLAN, Department of Social Policy, London School of Economics and Political Science
CAROL L. GRAHAM, The Brookings Institution, Washington, DC, and School of Public Policy, University of Maryland, College Park
V. JOSEPH HOTZ, Department of Economics, Duke University
DANIEL KAHNEMAN, Woodrow Wilson School of Public and International Affairs, Princeton University
ARIE KAPTEYN, Center for Economic and Social Research, University of Southern California, and RAND Corporation
AMANDA SACKER, Research Department of Epidemiology and Public Health, University College London
NORBERT SCHWARZ, Department of Psychology, University of Michigan
JUSTIN WOLFERS, Gerald R. Ford School of Public Policy, University of Michigan

CHRISTOPHER MACKIE, *Study Director*
ANTHONY S. MANN, *Program Coordinator*

COMMITTEE ON NATIONAL STATISTICS
2013-2014

LAWRENCE D. BROWN (*Chair*), Department of Statistics, The Wharton School, University of Pennsylvania
JOHN M. ABOWD, School of Industrial and Labor Relations, Cornell University
MARY ELLEN BOCK, Department of Statistics, Purdue University
DAVID CARD, Department of Economics, University of California, Berkeley
ALICIA CARRIQUIRY, Department of Statistics, Iowa State University
MICHAEL E. CHERNEW, Department of Health Care Policy, Harvard Medical School
CONSTANTINE GATSONIS, Center for Statistical Sciences, Brown University
JAMES S. HOUSE, Survey Research Center, Institute for Social Research, University of Michigan
MICHAEL HOUT, Department of Sociology, New York University
SALLIE ANN KELLER, Virginia Bioinformatics Institute at Virginia Tech, Arlington, Virginia
LISA LYNCH, The Heller School for Social Policy and Management, Brandeis University
COLM A. O'MUIRCHEARTAIGH, Harris Graduate School of Public Policy Studies, University of Chicago
RUTH D. PETERSON, Criminal Justice Research Center, Ohio State University
EDWARD H. SHORTLIFFE, Department of Biomedical Informatics, Columbia University and Arizona State University
HAL STERN, Donald Bren School of Information and Computer Sciences, University of California, Irvine

CONSTANCE F. CITRO, *Director*
JACQUELINE R. SOVDE, *Program Associate*

Acknowledgments

This report is the product of contributions from many colleagues, whom we thank for their insights and counsel. The project was sponsored by the National Institute on Aging (NIA) of the National Institutes of Health and by the UK Economic and Social Research Council (ESRC). We thank Richard Suzman and Lis Nielsen at NIA and Paul Boyle, Joy Todd, Ruth Lee, and Margot Walker at ESRC for their leadership in the area of subjective well-being (SWB) measurement and for their guidance and input to the project.

The panel also thanks the following individuals who attended the panel's open meetings and generously presented material to inform our deliberations. Angus Deaton (Princeton University) informed the panel about his analyses of Gallup data and other relevant research; Robert Groves (then director of the U.S. Census Bureau, now provost of Georgetown University) provided an overview of the potential role of federal surveys and statistical programs for advancing the measurement of SWB; and Richard Frank (Harvard University) and Jennifer Madans (National Center for Health Statistics of the Centers for Disease Control and Prevention) outlined the role of SWB measures in health research and policy and informed the panel about government experiences with them. Paul Allin, Stephen Hicks, Glenn Everett, and Dawn Snape (UK Office for National Statistics) provided overviews of exciting experimental work ongoing in the United Kingdom. Conal Smith, Carrie Exton, and Marco Mira d'Ercole (OECD) kept the panel abreast of their impressive work on the OECD *Guidelines on Measuring Subjective Well-being,* which was being conducted as the panel's work was under way. Somnath Chatterji (World Health Organization) discussed the organization's ongoing work on SWB as it pertains to health; Rachel Kranz-Kent (Bureau of Labor Statistics) provided an overview and plans for the agency's American Time Use Survey module on SWB; Michael Wolfson (University of Ottawa; formerly, Statistics Canada) informed the panel about Canada's experiences in developing and using well-being and quality-of-life measures; Steven Landefeld (Bureau of Economic Analysis) outlined the role of national economic accounts in measuring welfare and their relationship to measures of well-being; Michael Horrigan (Bureau of Labor Statistics) described his agency's interests in time-use statistics and well-being measures; Hermann Habermann (formerly with the U.S. Office of Management and Budget, the United Nations Statistical Division, and the U.S. Census Bureau) provided insights into U.S. and international statistical agencies' perspectives on the measurement of SWB; and Georgios Kavetsos and Laura Kudrna (London School of Economics) summarized their research findings (with panel member Paul Dolan) from analyses of data from the UK Office for National Statistics.

The panel could not have conducted its work efficiently without a very capable staff. Constance Citro, director of the Committee on National Statistics, and Robert Hauser, director of the Division of Behavioral and Social Sciences and Education (DBASSE), provided institutional leadership and substantive contributions during meetings; Kirsten Sampson-Snyder, DBASSE, expertly coordinated the review process; and Robert Katt provided thoughtful and thorough final editing. We also thank program coordinator Anthony Mann for his terrific logistical support of our local and overseas meetings.

On behalf of the panel, I especially thank the study director, Christopher Mackie, for his superb oversight of the panel's activities and his substantive contributions to the panel's work and this report. He skillfully and intelligently organized meetings and helped create a cordial and stimulating environment for conducting the panel's work. Chris mastered an entirely new domain of knowledge and contributed to the report by his careful and insightful editing of panel members' preliminary drafts of materials and diligent work on the final draft. And I would like to extend a personal note of gratitude to Chris for his unwavering optimism and good humor throughout this process; it was a delightful experience working with him on this project.

Most importantly, I would like to thank panel members for their patience, creativity, hard work, and graciousness when dealing with one another. Psychologists, sociologists, and economists often have different world views, and the panel was exceptionally cordial and considerate of all viewpoints. The report reflects collective expertise and commitment of all panel members: Norman Bradburn, University of Chicago; Laura Carstensen, Stanford University; Edward Diener, University of Illinois at Urbana-Champaign; Paul Dolan, London School of Economics and Political Science; Carol Graham, The Brookings Institution and University of Maryland, College Park; V. Joseph Hotz, Duke University; Daniel Kahneman, Princeton University; Arie Kapteyn, Center for Economic and Social Research, University of Southern California and RAND Corporation; Amanda Sacker, University College London; Norbert Schwarz, University of Michigan; and Justin Wolfers, University of Michigan. We all benefited from and enjoyed the depth of knowledge the panel members brought—literally—to the table.

This report has been reviewed in draft form by individuals chosen for their diverse perspectives and technical expertise, in accordance with procedures approved by the Report Review Committee of the National Research Council (NRC). The purpose of this independent review is to provide candid and critical comments that assist the institution in making its reports as sound as possible, and to ensure that the reports meet institutional standards for objectivity, evidence, and responsiveness to the study charge. The review comments and draft manuscript remain confidential to protect the integrity of the deliberative process.

The panel thanks the following individuals for their review of this report: Linda M. Bartoshuk, Center for Smell and Taste, University of Florida; Cynthia M. Beall, Department of Anthropology, Case Western Reserve University; Jennie E. Brand, Department of Sociology and California Center for Population Research, University of California, Los Angeles; Dora Costa, Department of Economics, Massachusetts Institute of Technology; Richard A. Easterlin, Department of Economics, University of Southern California; Jim Harter, Workplace Management and Wellbeing, Gallup; Martin Seligman, Department of Psychology, University of Pennsylvania; Dylan Smith, Center for Medical Humanities, Compassionate Care, and Bioethics, Stony Brook University; Jacqui Smith, Department of Psychology, University of Michigan; Tom W. Smith, NORC at the University of Chicago; Frank Stafford, Department of Economics, University of Michigan; Andrew Steptoe, Institute of Epidemiology and Health Care, University College London; and Joseph E. Stiglitz, Graduate School of Business, Columbia University.

Although the reviewers listed above provided many constructive comments and suggestions, they were not asked to endorse the conclusions or recommendations, nor did they see the final draft of the report before its release. The review of the report was overseen by James S. House, Survey

Research Center, Institute of Social Research, University of Michigan, and Ronald Brookmeyer, Department of Biostatistics, University of California, Los Angeles. Appointed by the NRC's Report Review Committee, they were responsible for making certain that the independent examination of this report was carried out in accordance with institutional procedures and that all review comments were carefully considered. Responsibility for the final content of the report rests entirely with the authoring panel and the NRC.

>Arthur A. Stone, *Chair*
>Panel on Measuring Subjective Well-Being
>in a Policy-Relevant Framework

Contents

SUMMARY	1
1 INTRODUCTION	15

   1.1 Overview of Subjective Well-Being, 15
      1.1.1 Evaluative Well-Being, 16
      1.1.2 Experienced Well-Being, 17
      1.1.3 Eudaimonic Well-Being, 19
   1.2 Study Charge, 20
   1.3 Motivation for Study, 21
   1.4 Report Audience, Report Structure, 26

2 CONCEPTUALIZING EXPERIENCED (OR HEDONIC) WELL-BEING   29
   2.1 Distinctiveness of Experienced and Evaluative Well-Being, 30
   2.2 Dimensions of ExWB, 36
      2.2.1 Negative and Positive Experiences—Selecting Content for Surveys, 36
      2.2.2 Eudaimonia, 40
      2.2.3 Other Candidate Emotions and Sensations for Measures of ExWB, 44

3 MEASURING EXPERIENCED WELL-BEING 49
   3.1 Ecological Momentary Assessment, 49
   3.2 Single-Day Measures, 52
      3.2.1 End-of-Day Measures, 52
      3.2.2 Global-Yesterday Measures, 54
      3.2.3 Appropriateness and Reliability of Single-Day Assessments of ExWB, 55
   3.3 Reconstructed Activity-Based Measures, 59
      3.3.1 Comparing DRM with Momentary Approaches, 61
      3.3.2 Time-Use Surveys, 66

4 ADDITIONAL CONCEPTUAL AND MEASUREMENT ISSUES 69
   4.1 Cultural Considerations, 69
   4.2 Aging and the Positivity Effect, 71
   4.3 Sensitivity of ExWB Measures to Changing Conditions, 72
   4.4 Adaptation, Response Shift, and the Validity of ExWB Measures, 75
   4.5 Survey Contextual Influences, 79
   4.6 Question-Order Effects, 81
   4.7 Scale Effects, 83
   4.8 Survey-Mode Effects, 84

5 SUBJECTIVE WELL-BEING AND POLICY 87
   5.1 What Do SWB Constructs Predict?, 91
   5.2 What Questions Can Be Informed by SWB Data: Evaluating Their Uses, 95
      5.2.1 The Health Domain, 95
      5.2.2 Applications Beyond the Health Domain, 98

6 DATA COLLECTION STRATEGIES 103
   6.1 Overall Approach, 103
      6.1.1 The Measurement Ideal, 105
      6.1.2 Next Steps and Practical Considerations, 109
   6.2 How to Leverage and Coordinate Existing Data Sources, 112
      6.2.1 SWB in Health Surveys and Other Special-Purpose Surveys, 113
      6.2.2 Taking Advantage of ATUS, 116
   6.3 Research and Experimentation—The Role of Smaller-Scale Studies, Nonsurvey Data, and New Technologies, 120

REFERENCES 125

APPENDIXES

A   Experienced Well-Being Questions and Modules from
    Existing Surveys   137
B   *The Subjective Well-Being Module of the American Time Use
    Survey: Assessment for Its Continuation*   153
C   Biographical Sketches of Panel Members   183

Summary

Research on subjective well-being (SWB), which refers to how people experience and evaluate their lives and specific domains and activities in their lives, has been ongoing for decades, providing new information about the human condition. During the past decade, interest in the topic among policy makers, national statistical offices, academic researchers, the media, and the public has increased markedly because of its potential for shedding light on the economic, social, and health conditions of populations and for informing policy decisions across these domains.

An impetus to this movement came from the 2009 report of the Commission on the Measurement of Economic Performance and Social Progress (Stiglitz et al., 2009), which concluded that government population surveys should be oriented toward measuring people's well-being, including the subjective dimension, as a way of assessing societal progress. The report emphasized that traditional market-based measures alone do not provide an adequate portrayal of quality of life, and recommended shifting the focus of economic measurement from production toward people's well-being. The underlying argument is that national policies should better balance growth in market production with considerations of equality, sustainability, and nonmarket dimensions of well-being that cannot be captured well by conventional "objective" measures.

Reflecting this interest in broadening and deepening the research base on SWB, the U.S. National Institute on Aging and the UK Economic and Social Research Council asked the National Research Council's Committee on National Statistics to convene an expert panel to (as described in the panel charge):

review the current state of research and evaluate methods for the measurement of subjective well-being (SWB) in population surveys . . . offer guidance about adopting SWB measures in official government surveys to inform social and economic policies . . . [and] consider whether research has advanced to a point which warrants the federal government collecting data that allow aspects of the population's SWB to be tracked and associated with changing conditions. . . . The study will focus on experienced well-being (e.g., reports of momentary positive and rewarding, or negative and distressing, states) and time-based approaches. . . . The connections between experienced well-being and life-evaluative measures will also be considered.

It should be made explicit that the panel's interpretation of its charge was to provide guidance primarily for measurement and data collection in the area of *experienced (hedonic) well-being* (ExWB). While acknowledging that measurement of the multiple dimensions of SWB is essential to a full understanding of it, this focus reflects the status of research on ExWB, which is less developed than it is for evaluative well-being, another dimension of SWB. Crucially, ExWB taps somewhat different domains of psychological functioning than do measures of evaluative well-being such as those dealing with life satisfaction. Indeed, many policy concerns—for example, those related to an aging population—center around quality of life and reduction of suffering on a day-to-day basis.

SWB data have already proven valuable to researchers, who have produced insights about the emotional states and experiences of people belonging to different groups, engaged in different activities, at different points in the life course, and living day to day in different family and community structures. Research has also revealed relationships between people's self-reported, subjectively assessed states and their behavior and decisions. In the broadest sense, the promise of information about people's ExWB rests in its capacity to enhance measures of (1) negative experiences, particularly where they can be shown to relate to longer-term suffering of specific populations in a way that provides insights into ways to reduce them, and (2) positive experiences, in a way that informs efforts to increase or enhance them. A reasonable analogy can be drawn with poverty. The policy goal in the 1960s in the United States to reduce poverty created the need to define and measure it. These tasks have been challenging, but knowledge generated by the process—although still incomplete, even after 50 years of effort—has proven essential for policy development and assessment. In the case of SWB, if, for example, long-term unemployment, depression, or lack of income are shown to be drivers of long-term suffering, then a strong case can be made for the inclusions in one or more datasets of such measures alongside information on employment status, mental health, and income.

In this report, a range of potential ExWB data applications are cited, from cost-benefit studies of health care delivery to commuting and transportation planning, environmental valuation, outdoor recreation resource monitoring, and assessment of end-of-life treatment options. Whether used to assess the consequence of people's situations and policies that might affect them or to explore determinants of outcomes (the impact of positive emotional states on resistance to or ability to recover from illness is now an actively researched example of the latter), contextual and covariate data are needed alongside the SWB measures.

DEFINING AND CHARACTERIZING ExWB

SWB is multifaceted and, for it to be a useful analytic construct, its components must be disentangled and understood. *Evaluative* well-being refers to judgments of how satisfying one's life is; these judgments are sometimes applied to specific aspects of life, such as relationships, community, health, and work. *Experienced* well-being—the focus of this report—is concerned with people's emotional states and may also include effects associated with sensations (e.g., pain, arousal) and other factors such as feelings of purpose or pointlessness that may be closely associated with emotional states and assessments of those states. The term "hedonic" is typically used to denote the narrower, emotional component of ExWB, which can be measured as a series of momentary states that take place through time. ExWB is often further divided into positive experiences, which may be characterized by terms such as joy, contentment, and happiness, and negative experiences, which may be characterized by sadness, stress, worry, pain, or suffering.

In some ways separate but also intertwined with the evaluative and experienced dimensions is *eudaimonic* well-being, which refers to a person's perceptions of meaningfulness, sense of purpose, and the value of his or her life. For thinking about some questions—such as the worthwhileness of specific activities, or the role of purpose in assessments of overall satisfaction with life—eudaimonic sentiments may be relevant to both experienced and evaluative measures of well-being. The most common assessment of eudaimonia refers to individuals' overall assessments of meaning and purpose.

In practice, a number of ExWB measurement approaches and objectives coexist, ranging from the moment-to-moment assessments of emotional states to questionnaires and interviews that require reflection by respondents about somewhat longer time periods, such as a whole day. ExWB measures can, in a sense, be viewed as a subspectrum of the overall SWB continuum that at one end involves a point-in-time reference period and is purely hedonic ("How do you feel at this moment?") and, at the other, involves an unstated but presumably much longer temporal reference period that is evaluative ("Taking all things together, how happy are you?"). As used in this report, ExWB includes the portion of the spectrum ranging from reports about feelings at a given moment to global-day assessments or reconstructions. Specification of the reference period has a strong impact on what will ultimately be measured. As the reference and recall periods lengthen, SWB measures take on more characteristics of life evaluation.

Unfortunately, in the literature, these temporal distinctions have often been blurred, which has led to ambiguous terminology and other confusions. "Happiness" has been used in reference to momentary emotional states and also as a way of describing overall life evaluations; such lack of specificity has at times muddled the discourse. Moreover, the multiple dimensions of well-being, such as suffering, pain, stress, contentment, excitement, purpose, and many others, cannot be ignored if investigators are to have any hope of understanding the complexities known to coexist. For example, a person who is engaged in stressful or difficult activities, such as working toward an education or a job promotion, may at the same time more broadly find meaning or satisfaction with life overall. Or a person who is chronically suffering or lacking hope may experience temporary reprieve in an enjoyable moment.

> **CONCLUSION 2.1:** Although life evaluation, positive experience, and negative experience are not completely separable—they correlate to some extent—there is strong evidence that multiple dimensions of SWB coexist. ExWB is distinctive enough from overall life evaluation to warrant pursuing it as a separate element in surveys; their level of independence demands that they be assessed as distinct dimensions.

The ExWB dimension of SWB itself can and often needs to be parsed more finely.

> **CONCLUSION 2.3:** Both positive and negative emotions must be accounted for in ExWB measurement, as research shows that they do not simply move in an inverse way. For example, an activity may produce both negative and positive feelings in a person, or certain individuals may be predisposed to experience both positives and negatives more strongly. Therefore, assessments of ExWB should include both positive and negative dimensions in order for meaningful inferences to be drawn.

Additionally, the observation that the aspects of negative experience—sadness, worry, stress, anger, etc.—tend to be more differentiated than those on the positive side, which are more unidimensional, carries implications for data collection.

SUMMARY

> **RECOMMENDATION 2.1:** When more than two ExWB questions can be accommodated on a survey, it is important to include additional ones that differentiate among negative emotions because—relative to the positive side—they are more complex and do not track in parallel (as the positive emotion questions tend to do).

At this point, empirical evidence does not indicate whether either the positive or negative ExWB dimension is more relevant to policy. But, as described in Chapter 5 and elsewhere in the report, reducing negative experiences, particularly those linked to prolonged suffering, is often a central policy objective, even if the exact levers have not been identified. To this end, development of a scale of "suffering" that has a duration dimension should be a pressing research concern. Such a measure might capture and distinguish between things like minutes of pain or stress versus ongoing poverty, hunger, and so on. Suffering is not the complete absence of happiness or the presence of exclusively negative experiences and emotions, and the scale should reflect this in a way that suggests relevant classes of policies.

To answer some kinds of questions, additional nuances beyond the positive and negative distinction are required. Thinking in terms of "experiences," as opposed to only "emotions," allows for consideration of an expanded set of factors—such as sense of purpose, hostility, pain, and others—which may also be important to developing a full picture of well-being.

> **CONCLUSION 2.4:** An important part of people's experiences may be overlooked if concepts associated with purpose and purposelessness are not included alongside hedonic ones like pleasure and pain in measures of ExWB. Crucially, central drivers of behavior may also go missing. People do many things because they are deemed purposeful or worthwhile, even if they are not especially pleasurable (e.g., reading the same story over and over again to a child, visiting a sick friend, or volunteering); they also do many things that are pleasant even if they are not viewed as having much long-term meaning in the imagined future.

When to include factors beyond the hedonic core depends on the research or policy question. For example, in studies of housing conditions or medical treatment effectiveness, sensations such as physical pain, numbness, heat, or cold, which enhance or degrade momentary experience, have an obvious relevance.

MEASURING ExWB

A range of techniques is available for measuring ExWB. At the short recall period end of the temporal spectrum are approaches that register emotional states in the moment.

> CONCLUSION 3.1: Momentary assessment methods are often regarded as the gold standard for capturing experiential states. However, these methods have not typically been practical for general population surveys because they involve highly intensive methods that are difficult to scale up to the level of nationally representative surveys and involve considerable respondent burden, which can lead to low response rates. For these reasons, while momentary assessment methods have proven important in research, they have not typically been in the purview of federal statistical agencies.

This conclusion reflects the current (and past) state of technology. The ways in which government agencies administer surveys is changing rapidly and, as monitoring technologies continue to evolve, new measurement opportunities will arise. For many, it may be less intrusive and burdensome to respond to a prompt from a programmed smartphone designed to sample real-time experience than to fill out a traditional survey. Use of such modes will become increasingly feasible, even for large-scale surveys, at reasonable cost.

The most frequently used alternatives, or compromises, to momentary assessment instruments are single-day measures, which involve questions asked at the end of the day or the day following the reference period (that is, about yesterday). Single-day measures have been shown to yield credible though somewhat different kinds of information about people's daily experiences. End-of-day methods, typically used in smaller-scale studies, cannot work with surveys that rely on interviews administered throughout the day. Given these constraints for momentary assessment and end-of-day approaches, global-yesterday questions have most often been used in large surveys.

> CONCLUSION 3.2: Global-yesterday measures represent a practical methodology for use in large population surveys. Data from such surveys have yielded important insights—for example, about the relationships between ExWB and income, age, health status, employment status, and other social and demographic characteristics. Research using these data has also revealed how these relationships differ from those associated with measures of evaluative well-being. Even so, there is much still to be learned about single-day measures, and it is pos-

sible that much of what has been concluded so far may end up being contested.

For some research and policy questions, contextual information about activities, specific behaviors, and proximate determinants is essential. For example, if the question is how people feel during job search activities, while undergoing medical procedures, or engaged in child care, more detailed information than can be typically ascertained from a global daily assessment is needed. Activity-based or time-use methods—such as the Day Reconstruction Method (DRM)—attempt to fill this measurement need. The DRM asks respondents to describe the day's events by type of activity (e.g., commuting to work, having a meal, exercising) and provide a detailed rating of their emotional state during the activity. The DRM therefore goes beyond asking who is happy to asking when they are happy. This time-use dimension potentially establishes links to policy levers.

CONCLUSION 3.6: Capturing the time-use and activity details of survey respondents enhances the policy relevance of ExWB measures by embedding information about relationships between emotional states and specific activities of daily life.

The nature of the question under consideration dictates the appropriate measurement method and may suggest an appropriate data collection modality. For example, if the particular SWB dimension of interest is thought to be sensitive on a short time frame—to daily activities (e.g., going for a run) or events (e.g., a big win by one's favorite team)—a large cross-sectional data collection conducted every 2 years is unlikely to be useful. In such cases a high-frequency approach (even if it involves much smaller samples) might be more informative, and less costly.

ADDITIONAL MEASUREMENT ISSUES

This report addresses a number of conceptual and survey methodology issues pertaining to SWB measurement; among the most crucial are

Sensitivity of measures to changing conditions, situations. A prerequisite to applying SWB data to policy is understanding what constitutes a meaningful change in a measure. In thinking about "sensitivity" and how measures are calibrated, it is instructive to consider standards applied to existing statistics. A change in the unemployment rate, for example, from 6 percent to 6.1 percent reflects a change in status of only 1 in 1,000 people in the workforce. Over the 50 years that the unemployment statistic has existed, analysts have had time to learn how to interpret what appears to be a small change; key here is that a 0.1 percent change in the unemployment rate represents a much larger impact among the population defined as actively looking for work than for the total workforce. Similarly, it will take time to understand how to interpret SWB time series data.

Survey context, ordering, and mode effects. Although a survey methodology concern generally, question ordering and contextual factors appear to be especially serious for subjective well-being. An experimental split-sample randomized trial conducted by the UK Office for National Statistics (ONS) reported a significant question-order effect for multiple-item positive and negative affect questions: it mattered whether the positive questions or the negative questions were answered first. Deaton's (2012) analysis of Gallup-Healthways data demonstrated the importance of the type of questions (and responses) that precede well-being assessments. Specifically, asking questions about political topics first had a substantial impact on a subsequent measure of evaluative well-being, though it had relatively little effect on the ExWB measure. Insertion of buffer questions has in some cases been shown to virtually eliminate item-order effects, suggesting that careful survey design has the potential to greatly minimize these problems. As noted in section 4.6, many of these design questions can be addressed using fairly straightforward experiments that will ultimately lead to better surveys.

Survey mode refers to how questions are posed to respondents—for example, by personal interview, telephone, or Internet instrument. Results from another split sample of the ONS survey found significantly higher life-satisfaction, happiness, and worthwhile scores (and lower anxiety scores) for telephone interviews compared to face-to face interviews, suggesting that survey mode can have a significant impact on respondent ratings.

RECOMMENDATION 4.3: Given the potential magnitude of survey-mode and contextual effects (as shown in findings related to work by the UK Office for National Statistics and elsewhere), research on the magnitude of these effects and methods for mitigating them should be a priority for statistical agencies during the process of experimentation and testing of new SWB modules.

Another important methodological issue that has arisen in the literature, discussed in sections 4.1 and 4.2, is whether respondents' answers to SWB questions are subject to biases among groups—defined by culture, age, or other traits—that may invite misleading conclusions about actual experiences. Research has shown systematic variations in reported well-being that appear to be associated with cultural norms about ideal emotional states. Another potential threat to the validity of ExWB measures, discussed in section 4.4, is adaptation: the psychological process whereby people adjust to and become accustomed to a positive or negative stimulus brought on by changed circumstances.

POLICY RELEVANCE

A major research challenge is to improve the knowledge base about causal pathways—both between SWB and its determinants and between SWB and various outcomes—in a way that would be suggestive of policy mechanisms. Understanding causal properties is, of course, a difficult problem in many areas of social science, not just for research on SWB. Heckman (2000, p. 91) has aptly described this general difficulty for his own discipline:

> Some of the disagreement that arises in interpreting a given body of data is intrinsic to the field of economics because of the conditional nature of causal knowledge. The information in any body of data is usually too weak to eliminate competing causal explanations of the same phenomenon. There is no mechanical algorithm for producing a set of "assumption free" facts or causal estimates based on those facts.

This critique seems especially pertinent for analyses of SWB data. In many situations, it is not known whether positive and negative emotions are the predictor or outcome or if the association is reciprocal. For example, the observed association between positive emotional states and better health may be causally linked in that order, or better health may create conditions for happiness. Clearly, both can be taking place. Income and well-being could also embody this kind of circular interaction, in which distinguishing cause and effect is difficult.

The unique policy value of ExWB measures may not be in new assessments of how income does or does not relate to SWB or in an aggregate-level tracking of experiential states. Rather, their value may come from the discovery of actionable relationships for specific policies—in such diverse areas as health, city planning and neighborhood amenities, divorce and child care practices and laws, commuting infrastructure, recreation and exercise, social connectedness, and corruption—that may otherwise escape attention.

CONCLUSION 5.1: ExWB data are most relevant and valuable for informing specific, targeted policy questions, as opposed to general monitoring purposes. At this time, the panel is skeptical about the usefulness of an aggregate measure intended to track some average of an entire population.

Perhaps the most compelling reason for pursuing ExWB data collection is its potential to identify subpopulations that are suffering and to inform research into the sources of and solutions to that suffering. Again, the panel emphasizes the necessity of measuring both experienced and evaluative dimensions of self-reported well-being. Certain policies may aim to enhance one or the other of these dimensions but may end up affecting both. For instance, an action designed to enhance day-to-day living quality at the end of life may have an impact on life satisfaction as well. And policies that aim to enhance longer-term opportunities of the young may in turn have short-term negative effects on momentary emotional experience—as in the case of a student who must work hard in school, which may at times be unpleasant, but pays off later in terms of higher life satisfaction.

CONCLUSION 5.2: To make well-informed policy decisions, data are needed on both ExWB and evaluative well-being. Considering only one or the other could lead to a distorted conception of the relationship between SWB and the issues it is capable of informing, a truncated basis for predicting peoples' behavior and choices, and ultimately compromised policy prescriptions.

DATA COLLECTION STRATEGIES

Because self-reported well-being embodies multiple dimensions and sheds light on behavior and conditions at different levels of aggregation, an ideal measurement infrastructure requires a multipronged approach.

The Measurement Ideal

One prong of a comprehensive SWB measurement program involves inclusion of modules in large-scale population surveys such as those in the ONS Integrated Household Survey and the Gallup World Poll. The repeated cross-sectional structure of such surveys allows both evaluative well-being and ExWB to be tracked. These sources are capable of identifying suffering or thriving subgroups, facilitating qualitative research for special populations, and perhaps providing useful policy information at the macro level. The Gallup data have also been used for nation-to-nation comparisons.

The second prong of a comprehensive measurement program involves inclusion of SWB questions in specialized, focused data collections. Examples include health interview surveys, time-use surveys, and neighborhood environment surveys. Question modules may be constructed as experiments or pilots within existing large survey programs (the American Time Use Survey [ATUS] module, for example, uses outgoing samples of the Current Population Survey), or they may stand alone, in which case they may be designed to include covariates shown or thought to have the strongest associations with ExWB. The advantage of targeted studies is that they can be tailored to address specific questions—whether about health care, city planning, or airport noise management—and can sometimes be attached to ongoing surveys for which the surrounding content is appropriate. Another example is the American Housing Survey's new Neighborhood Social Capital module; adding ExWB questions would allow researchers to explore links to community characteristics, connectedness, and resilience—associations specifically cited by Stiglitz et al. (2009) as very important and potentially alterable by policy. Because research continues to reveal details about the links between healthy emotional states and healthy physical states, health surveys provide an increasingly secure foothold for ExWB measurement. An appealing feature of smaller-scale or special-purpose surveys is that they can often be supported by funding agencies in such a way that content matches well with their organizational missions.

The third prong to an ideal data infrastructure consists of panel data collection. Information about how individuals' SWB changes over time and in reaction to events and life circumstances cannot be fully understood without longitudinal information; such data are also crucial for addressing questions of causality (e.g., does getting married make people happier, or are happier people more likely to get married?). Krueger and Mueller (2012), for example, were able to examine the emotional impact associated with job search and other daily activities for the unemployed, both during joblessness and upon reemployment, using longitudinal time-use data. Just as panel data have allowed researchers to learn more about the characteristics of poverty (revealing less chronic poverty and more movement in and out of poverty than was once thought), panel data on ExWB may be useful to researchers studying the duration of depression and suffering at the individual level and whether these conditions tend to be chronic or if there is movement in and out of suffering states and groups. It is difficult to study such phenomena without panel data that are collected on a regular and frequent basis.

A final component of an ideal ExWB data collection strategy is real-time data collection. As described above, momentary sampling methods have been central to ExWB research but are often impractical for national statistical offices. For the immediate future, the primary means for measuring and tracking ExWB, and SWB more broadly, will continue to be survey based. Neither the technical or economic challenges to "traditional" survey methods nor the promises of alternative ways for measuring the public's behaviors and views have reached a point where it is sensible to transition away completely from the former.

However, although real-time, momentary monitoring may not now be practical for major surveys such as the American Community Survey or the Current Population Survey, it may be (or become) a reality for a number of other surveys, particularly in the health realm. Knowing how people are feeling and what they are doing at the same moment can shed light on the relationships between ExWB and a long list of correlates from commuting, to air pollution, to child care, with clear ties to policy. As the ways in which government agencies administer surveys change—in reaction not just to rapidly evolving technology but also to declining response rates and escalating survey costs—new measurement opportunities will arise. For government data collection to stay relevant and feasible, statistical agencies will need to apportion some of their resources to understanding and adapting to emerging survey methods, new "big data" sources, and alternative computational science methods for measuring people's behavior, attitudes, and states of well-being.

Assessment of Current Data Collection

Very few if any national statistical offices have the resources needed to pursue data collection on all the fronts identified above as parts of the ideal strategy. At this point, some data collection modes are better understood and better supported by evidence linking them to outcomes than others; phasing in SWB data collection should reflect this. While recent research has rapidly advanced our understanding of the properties of SWB measures and their determinants, ExWB metrics are not yet ready to be published and presented as "official statistics."

> **RECOMMENDATION 6.1:** ExWB measurement should, at this point, still be pursued in experimental survey modules. The panel encourages inclusion of ExWB questions in a wide range of surveys so that the properties of data generated by them can be studied further; at this time, ExWB questions should only be considered for inclusion in flagship surveys on a piloted basis. Numerous unresolved methodological issues such as mode and question-order effects, question wording, and interpretation of response biases need to be better understood before a module should be considered for implementation on a permanent basis.

The United Kingdom, because of its more centralized statistical system and the opportunity raised by the current government's interest in well-being measurement, has been able to push further than has the United States on the first prong of the comprehensive measurement infrastructure laid out above. The cautions noted above notwithstanding, it is important to recognize and commend the opportunity that the ONS initiative has provided to begin analyzing data properties, interpreting the results, and generally using it as a test bed for further development of SWB measurement. However, for the United States, the panel recommends prioritizing development of SWB modules for inclusion in targeted, specialized surveys above the development of instruments for the large general population surveys.

> RECOMMENDATION 6.2: ExWB questions or modules should be included (or should continue to be included) in surveys where a strong case for subject-matter relevance can be made—those used to address targeted questions where SWB links have been well researched and where plausible associations to important outcomes can be tested. Good candidates include the Survey of Income and Program Participation (which offers income, program participation, and care-giver links); the Health and Retirement Study (health, aging, and work transition links); the American Housing Survey's Neighborhood Social Capital module (community amenities and social connectedness links); the Panel Study of Income Dynamics (care-giving arrangements, connectedness, and health links); the National Longitudinal Survey of Youth (understanding patterns of obesity); and the National Health Interview Survey and the National Health and Nutrition Examination Survey (health and health care links).

The ATUS modified DRM module is the most important U.S. government ExWB data collection, and its continuation would enhance SWB research. It also provides an appropriate vehicle for experiments to improve the structure of abbreviated DRM-type surveys. The ATUS SWB module is the only federal government data source of its kind—linking self-reported information on individuals' well-being to their activities and time use. Though there are no plans to field it in 2014 (or beyond) at this point, the SWB module is practical, inexpensive, and worth continuing as a component of ATUS. Not only does the ATUS SWB module support research, it also provides additional information to help refine SWB measures that may one day be added to the body of official statistics.

1

Introduction

1.1 OVERVIEW OF SUBJECTIVE WELL-BEING

Subjective well-being (SWB) refers to how people *experience* and *evaluate* their lives and specific domains and activities in their lives. Over the past decade, interest in information about SWB (also called "self-reported well-being") has increased markedly among researchers, politicians, national statistical offices, the media, and the public.[1] The value of this information lies in its potential contribution to monitoring the economic, social, and health conditions of populations and in potentially informing policy decisions across these domains (Krueger et al., 2009; Layard, 2006).

Economists, psychologists, and sociologists have found a number of distinct components of SWB to coexist but which are not entirely independent—they do overlap. These measurement constructs may be thought of in terms of a continuum, with essentially real-time assessments of experience, emotional state, or sensations at one end (associated with the shortest time unit) and overall evaluations of life satisfaction, purpose, or suffering at the other end (the longest reference periods or no particular reference period).

These temporal overlaps notwithstanding, the components of SWB display distinct characteristics, often correlate with different sets of variables, and capture unique aspects of the construct that for various purposes are each worth monitoring. The terms used to describe SWB have often been

---

[1] OECD (2013) notes that, just in economics, a search of the *Econlit* database for a recent year (2008 is cited) returns more than 50 articles per year on SWB whereas, for the 1990s, the same search returns fewer than 5 per year, on average.

ambiguously applied, which has muddled discussion and possibly slowed progress in the field. For example, the term "happiness" has been used to refer to momentary assessments of affect as well as to overall life evaluations. This absence of precision precludes understanding of the complexities known to coexist. For example, a person who is engaged in stressful or difficult activities, such as working toward an education or a job promotion, may find substantial meaning or satisfaction with life overall; a person who is generally suffering or lacking hope may experience temporary reprieve in an enjoyable moment.

The nature of the policy or research question being asked dictates the appropriate construct to measure SWB and may suggest an approach to data collection. For example, if the dimension of interest is known to be sensitive on a very short time frame and responds to daily activities and events but is somewhat stable over long periods, a cross-sectional data collection conducted every 2 years may not be useful. In such cases, a high-frequency approach (even if it involves a much smaller sample) might be most informative.[2] Similarly, if a measure varies a great deal from individual to individual on a given day but does not react very much to exogenous events (financial shocks, changes in employment rates, etc.) and tends to wash out at high aggregate levels, it may not be a particularly insightful construct to track at national levels over time.

The following sections briefly identify the distinct components that must be measured in order to produce a full and clear accounting of SWB. Chapter 2 discusses these components and the interactions among them in greater detail.

1.1.1 Evaluative Well-Being

Measures of evaluative well-being are designed to capture judgments of overall life satisfaction or fulfillment; these judgments may be applied to specific aspects of life, such as relationships, community, health, or occupation, as well as to overall evaluations. An example of a question phrased to measure evaluative well-being—one recommended by OECD (2013, p. 253) and based on the World Values Survey—is "Overall, how satisfied are you with life as a whole these days?" Although OECD has proposed a scale from 0 to 10 for this question (OECD, 2013, p. 254), different scales have been used for versions of the question by other surveys, including the UK Office for National Statistics (ONS), the French national statistics

---

[2] Consumer confidence, for example, can display this kind of pattern, which may be a reason that the survey on which the University of Michigan Consumer Sentiment Index is based is designed as it is—with fairly small samples but ongoing data collection.

office, the British Household Panel Study, the Canadian General Social Survey, the German Socioeconomic Panel, and the European Social Survey.

Alternative measures of evaluative well-being exist, such as the CASP-19, a quality-of-life scale for older people that is often used in research on aging (Hyde et al., 2003), the Cantril Self-Anchoring Striving Scale (Cantril, 1965), and the five-item scale designed by Diener et al. (1985) to measure global cognitive judgments of life satisfaction. The Cantril Scale is the instrument for measuring evaluative well-being used in several Gallup initiatives, including the World Poll.[3] Research (e.g., Fredrickson et al., 2013) suggests that different aspects of well-being may have distinct physiological correlates. Longitudinal studies indicate moderate stability of life satisfaction over time; the variation that has been observed suggests there are potentially modifiable contextual factors that influence judgments about some aspects of evaluative well-being.

1.1.2 Experienced Well-Being

Experienced well-being (ExWB)—the focus of this report—is closely related to the oft-used term "hedonic well-being,"[4] which Christodoulou et al. (2013, p. 2) characterized as referring to:

> the frequency and intensity of emotional experiences such as happiness, joy, stress, and worry that make a person's life pleasant or unpleasant (Kahneman and Deaton, 2010). A variety of disciplines have shown increasing interest in the accurate assessment of HWB [hedonic well-being], especially positive aspects of well-being (Seligman and Csikszentmihalyi, 2000; Kahneman and Krueger, 2006; Huppert et al., 2004; Krueger et al., 2009). Research has begun to delineate the neurobiological foundations of [hedonic well-being] (Davidson, 2004) and to discern broad and important implications in areas such as health and society. In health research, positive affect has been found to predict response to illness (Cohen et al., 2003) and even survival among older men and women (Steptoe and Wardle, 2011). In the economic and social arenas, there is a realization that traditional economic

---

[3] The Cantril Self-Anchoring Scale asks respondents to imagine a ladder with steps numbered from 0 at the bottom to 10 at the top, in which the top of the ladder represents the best possible life for them and the bottom of the ladder represents the worst possible life. They are asked which step of the ladder they personally feel they stand on at this time (for a present assessment). For a good description and discussion of the Cantril Scale, see Diener et al. (2009).

[4] The terms "hedonic well-being" and "experienced well-being" are often used interchangeably in the literature. Interpreted more precisely, the latter is a somewhat broader concept in that hedonic well-being refers specifically to moment-to-moment emotional states, while experienced well-being may be extended to include sensations (e.g., pain, arousal) or other factors beyond emotions. However, the two terms are very closely related, especially because the additional "experience" dimensions of the latter concept may directly impact the individual's emotional states.

measures such as income provide an incomplete explanation of societal well-being (Easterlin, 2001; Kahneman and Deaton, 2010) and that appropriate measurement of [hedonic well-being] could serve as a useful complement to traditional economic indicators (Kahneman et al., 2004; Seaford, 2011).

Thus, measures of ExWB are designed to reflect some combination of "positives," such as pleasure, joy, contentment, or happiness, and "negatives," such as suffering, distress, sadness, stress, or worry. These measures are obtained from personal (subjective) reports that are made either in real time or shortly after an event has occurred.

The distinction between positive and negative emotions (or affect) is essential, as evidence is conclusive that one is not simply the inverse of the other. And there is little doubt that positive and negative dimensions track at least partially independently of life satisfaction and of each other. Additionally, other dimensions of ExWB, such as anger or arousal, which relate to positive and negative emotions in a range of ways, are important. Sensations such as pain may also figure into emotional states and into hedonic assessment of those states. Finally, cognitive appraisals of the meaning, purpose, or worthwhileness of current activities may also be included in the ExWB construct.

Examples of techniques for measuring ExWB, discussed in detail in Chapter 3, include applications of the Positive and Negative Affect Schedule (Watson et al., 1988) and a range of approaches involving Ecological Momentary Assessment (Stone and Shiffman, 1994). Data on ExWB have been collected less frequently in large surveys than have data on life evaluations, and methods for collecting data on hedonic experience in real time—experience sampling—have rarely been applied to a representative population sample because they are burdensome. Less intense methods, such as the Day Reconstruction Method (Kahneman et al., 2004), designed to help individuals recover their experiences and associated emotions of the day before (described in detail below), have been implemented through representative samples. Another class of single-day measurement approaches for ExWB, such as that used in the Gallup-Healthways Well-Being Index, asks about the presence of a range of emotions the previous day; others ask about emotional experience at the end of the reference day.

Along with the life-evaluation questions, the OECD *Guidelines* recommend a global-yesterday question for use in a module designed to include a minimal set of measures for use in government household surveys. Derived from the Gallup World Poll and European Social Survey, the recommended questions are phrased as follows (OECD, 2013, p. 253):

> The following question asks about how you felt yesterday on a scale from 0 to 10. Zero means you did not experience the feeling "at all" yesterday

while 10 means you experienced the feeling "all of the time" yesterday. I will now read out a list of ways you might have felt yesterday.

A3. How about happy?
A4. How about worried?
A5. How about depressed?

Other surveys with components to measure ExWB use different (sometimes very different) emotion or affect adjectives.[5]

1.1.3 Eudaimonic Well-Being

Eudaimonic well-being refers to people's perceptions of the meaningfulness (or pointlessness), sense of purpose, and value of their life—a very broad set of considerations. The ancient Greek concept of *eudaimonia* implies a premise that people achieve happiness if they experience life purpose, challenges, and growth. "Flourishing" is a term that has been suggested (Keyes, 2002) as capturing the essence of this dimension of well-being. An example of a eudaimonic question—developed by ONS for the Annual Population Survey—asks respondents, "Overall, to what extent do you feel the things you do in your life are worthwhile?" In this case, a 0 to 10 scale is used, where 0 means the respondent feels the things they do in their life "are not at all worthwhile" and 10 means "completely worthwhile" (OECD, 2013, p. 253).

There has been less research into eudaimonic well-being than into either evaluative or ExWB; consequently, its role in explaining behavior is less well understood. For some questions, such as the "worthwhileness" of specific activities or the role of purpose in a person's assessment of overall satisfaction with life, eudaimonic sentiments may figure into emotional states or into evaluations of life satisfaction. All subjective reports involve either evaluations or experiences, or both. However, concepts of "worthwhileness" or purpose appear crucial for understanding (or predicting) why and when people engage in various activities during the day or choose various life courses. White and Dolan (2009) have measured the worthwhileness (reward) associated with activities using day reconstructions of time and activities. They find discrepancies between those activities that people find "pleasurable" as compared to "rewarding" or meaningful. For example, time spent with children is relatively more rewarding than pleasurable, whereas time spent watching television is relatively more pleasurable than rewarding.

---

[5] See Appendix A for a more extensive sample of questions currently in use to evaluate self-reported well-being.

1.2 STUDY CHARGE

The objective of this report is to:

- Review the current state of research and evaluate methods for measuring self-reported *hedonic* (or *experienced*) well-being that are useful for monitoring, informing, and policy analysis purposes. Although the emphasis of this report is on ExWB and time-based approaches, their relationships with measures of evaluative well-being are considered. The report does not assess the value of evaluative well-being measures.
- Assess whether research on, and the methods to study, ExWB have advanced to a point that warrants the federal government collecting data in surveys and constructing indicators, accounts, or other statistics to inform social and economic policies—recognizing that the UK and U.S. statistics agencies are at different stages of development with regard to measurement constructs for SWB and operate within very different systems. In assessing the reliability and value of data on ExWB, the point of comparison should be other measures that are routinely collected; otherwise the comparison may be with a perfect world, not the real one.
- Recommend strategies for implementing data collection on ExWB, or, if premature, outline work that needs to be done before moving measurement of ExWB to statistical agency agendas.

The panel charge, verbatim, is reproduced in Box 1-1.

The value of research on SWB and the insights it has produced have been well established in the literature over recent decades. Much of this research has relied on nongovernment data collections, such as those conducted by the Gallup Organization. A central task of this study is to assess and provide guidance about the optimal role that statistical agencies might play in collecting, coordinating, and publishing data needed to advance the field further and potentially to inform policy discussions.

It should be made explicit here that the panel's interpretation of its charge was to provide guidance primarily for the measurement and data collection in the area of *experienced (hedonic) well-being*. In line with this emphasis, this report partially sets aside a substantial body of work on policy-relevant measures of evaluative well-being.[6] Guides to this work may be found in the 2009 report by Stiglitz, Sen, and Fitoussi, the recently released OECD *Guidelines on Measuring Subjective Well-being* (OECD,

---

[6] We qualify with "partially" because the relationships between experienced and evaluative well-being are described in some detail in section 2.1.

> **BOX 1-1**
> **Panel Charge**
>
> An ad hoc panel will review the current state of research and evaluate methods for the measurement of subjective well-being (SWB) in population surveys. On the basis of this evaluation, the panel will offer guidance about adopting SWB measures in official government surveys to inform social and economic policies. The study will consider whether research has advanced to a point which warrants the federal government collecting data that allow aspects of the population's SWB to be tracked and associated with changing conditions.
>
> The study will focus on experienced well-being (ExWB) (e.g., reports of momentary positive and rewarding, or negative and distressing, states) and time-based approaches, some of the most promising of which are oriented toward monitoring misery and pain as opposed to "happiness"; however, the connections between ExWB and life-evaluative measures will also be considered. Although primarily focused on SWB measures for inclusion in U.S. government surveys, the panel will also consider inclusion of SWB measures in surveys in the United Kingdom and European Union, in order to facilitate cross-national comparisons in addition to comparisons over time and for population groups within the United States.
>
> The panel will prepare an interim report on the usefulness of the American Time Use Survey SWB module and a final report identifying potential indicators and offering recommendations for their measurement.

2013) or the *World Happiness Report* (Helliwell et al., 2012), to name just a few. Additionally, the sponsors of this study (the U.S. National Institute on Aging and the UK Economic and Social Research Council)—which are keenly interested in the development and refinement of measures and concepts covering the full range of well-being—have noted that the measures of well-being used in aging research have focused almost exclusively on life satisfaction addressing many questions, which they rightfully argue is not sufficient. Our understanding of ExWB is more incomplete, yet its measurement may be equally valuable in that it likely taps somewhat different domains of psychological functioning. Indeed many of the concerns related to an aging population center around quality of life, well-being, and the reduction of suffering on a day-to-day basis.

1.3 MOTIVATION FOR STUDY

Data collections on SWB and related constructs have already proven to be highly valuable to researchers, producing insights into the emotional states and self-evaluated life satisfaction of people belonging to different groups, engaged in different activities, at different points in the life course, and involved in different family and community structures. Research has also shown how these subjectively assessed states of individuals relate to their behavior and decisions. Additionally, the media, politicians, and the general public have shown a strong interest in the information portrayed in these data and statistics.

The case for policy relevance is still developing but is well in motion. In the broadest sense, the promise of studying self-reported well-being rests in its capacity to enhance measures of (1) suffering (particularly long-term suffering) in a way that provides insights into its reduction, and (2) positive experiences in a way that informs efforts to increase or enhance them. A reasonable analogy can be drawn with poverty. Once poverty reduction emerged as a policy priority, a need to define and measure it (i.e., to design a poverty measure) was created. And to be most useful, information needed to be simultaneously collected on variables, such as education, health, economic mobility, and other factors that relate to poverty, whether as a cause, as a result, or in a circular fashion. This analogy also highlights the need to embed measurement of SWB in the most useful contexts. For example, if long-term unemployment, depression, lack of income, or lack of social connectedness prove to be drivers of long-term suffering, appropriate datasets are those that include covariate information on employment status (e.g., Current Population Survey's American Time Use Survey [ATUS]), mental health (e.g., National Health Interview Survey), income (e.g., Survey of Income and Program Participation), and social capital (e.g., American Housing Survey's Neighborhood Social Capital module). Likewise, promising data collection vehicles would be implied if positive affect were shown to have a measurable impact on health or workforce productivity.

In Chapter 5, the panel cites several policy applications or potential applications, ranging from assessment of end-of-life treatment options, cost-benefit studies of health care delivery (particularly where dimensions not captured by longevity or quality-adjusted life year metrics are present), and commuting and transportation planning, to environmental valuation and outdoor recreation resource monitoring. Beyond cases where SWB data may allow for fuller cost-benefit analyses of policy options, there may also be reverse cases, where measures of people's SWB are indicative of a factor driving outcomes; the impact of positive affect on resistance to or ability to recover from illness is an actively researched example. As is true for most measures, even those viewed as "objective," the goal is not to have a perfect measure of SWB but to generate data that can be usefully combined with other information and incorporated in a range of policy applications.

Spurred by the types of questions described above—along with an increasing desire by policy makers, researchers, and the public for a richer concept of progress and well-being than can be provided by traditional market-based measures on their own—research on SWB has recently accelerated and calls for data collection by statistical offices have been invigorated. Pointed impetus to the movement was provided by the Commission on the Measurement of Economic Performance and Social Progress, established by French President Nicolas Sarkozy and chaired by Joseph Stiglitz, which argued that governments and population surveys should measure people's well-being as a way of assessing societal progress (Stiglitz et al., 2009). The Commission included a working group that analyzed new measures of quality of life, including subjective ones, and its report emphasized that economic growth alone (as measured, for example, by growth in gross domestic product [GDP]) is not a satisfactory measure of the standard of living. The Commission recommended a shift in the focus of economic measurement from production toward people's well-being (Stiglitz et al., 2009). The underlying argument is that, at least in developed nations, per capita GDP is high (some argue that societies now over-consume) and the focus of national policies should shift to issues of inequality (even with high per capita GDP, those at the bottom of the economic ladder still suffer), sustainability, and nonmarket dimensions of well-being that cannot all be well captured by conventional, "objective" measures of well-being.

Emerging and ongoing efforts around the world to establish measures of and statistics on SWB also provide a strong impetus for this report. Initiatives by national statistical offices and international organizations are very much in their experimental phases, so this is the time to contribute input to them. The panel sees a clear need emerging to provide guidance for next steps to advance data, surveys, and research on the subject. An overarching part of such guidance is the need for clarification and a better understanding of the different dimensions of SWB, the specific information added by data on measures of ExWB, and the kinds of policy-relevant questions such data would inform.

In the United States, ATUS has, since 2010, included a module asking respondents about feelings (pain, happiness, stress, sadness, tiredness) during specific episodes of the day. Given the extensive and rapidly growing academic literature on time use and SWB (see references at front of chapter), this is an appropriate time to assess that literature and to determine whether and how to apply it in the statistical policies of the U.S. government. Appendix B to this report provides support for and guidance on the continuation and development of the ATUS module on SWB.

Among efforts currently under way that are attempting to advance measurement of SWB among national statistical offices, perhaps the most prominent is the recently developed and published OECD *Guidelines on Measuring Subjective Well-being* (2013). The *Guidelines* are intended to:

Improve the quality of subjective well-being measures collected by national statistical offices, by providing best practice in terms of question wording and survey design; improve the usefulness of the data collected by setting out guidelines on the appropriate frequency, survey vehicles, and co-variates when collecting subjective well-being data; improve cross-country comparability of subjective well-being measures by establishing common concepts, classifications, and methods that national statistical agencies could use; and provide advice and assistance to data users when analyzing subjective well-being data. (OECD, 2013, p. 9)

National statistical offices are now being called upon to begin systematically gathering and publishing information on subjective measures of well-being. ONS now includes a set of four questions on the core of its Integrated Household Survey covering three aspects of SWB: life evaluation, momentary emotional state, and worthwhileness. Beginning in April 2011, ONS included the following questions on its Annual Population Survey and Opinions and Lifestyle Survey:

- Overall, how satisfied are you with your life nowadays? [evaluative well-being]
- Overall, to what extent do you feel the things you do in your life are worthwhile? [eudaimonic well-being]
- Overall, how happy did you feel yesterday? [experienced well-being]
- Overall, how anxious did you feel yesterday? [experienced well-being]

All were answered on a scale of 0 to 10 where 0 is "not at all" and 10 is "completely."[7]

Elsewhere, the French national statistical office has collected information on SWB, and on ExWB specifically, in the Enquete Emploi du Temps 2009-2010. Plans are in motion to collect data on SWB by the statistical systems in a number of other European nations and beyond, including South Korea and Japan. Chile now has a life satisfaction question in its annual National Socioeconomic Survey, which produces high-quality, annual poverty information at the household level. Other countries have long collected information on SWB: Canada has done so in the General Social Survey since 1985; New Zealand collects data on life satisfaction through its General Social Survey; and Australia has collected information on SWB in its Household, Income and Labour Dynamics in Australia Survey. In addition to several new initiatives, the Japanese government has collected data on SWB continuously since 1958 in its Life in Nation Surveys (Stevenson and Wolfers, 2008). Eurostat began developing a module on SWB for the

---

[7]For more information on the ONS program, see *Measuring Subjective Well-being in the UK* on the ONS website: http://www.ons.gov.uk/ons/index.html [October 2013].

European System of Social Surveys. Some international agencies, such as the World Health Organization, have long worked with quality-of-life measures; typically these have been assessments of evaluative, as opposed to experienced, well-being.

Although a few national statistical offices are in the forefront of obtaining regular measures of well-being, most (including the United States) have played only a limited role in this regard. Indeed, some of the most prominent surveys measuring SWB and comparing countries' performance are undertaken by commercial and academic organizations. The most widely used (and largest) datasets on SWB are the Gallup Organization's World Poll—begun in 2005 and covering 160 countries—and Gallup World Values Survey. The Gallup World Poll is a repeated annual cross-sectional survey that includes life evaluation and ExWB questions, as well as many factors beyond self-reported well-being, such as perceptions of work, social, financial, physical, and community well-being; perception of leadership; basic access to food, shelter, safety; and others. In 2008, Gallup instituted a daily poll of 1,000 individuals in the United States that includes evaluative and ExWB. The World Values Survey, which is also cross-sectional, collects information on life evaluation and overall happiness and has sometimes also included questions asking about more focused measures of experienced emotion and mood.

A number of national surveys conducted by academic institutions, often funded by governmental organizations, include assessments of SWB as a component of their standard questionnaire/interview protocol. These are usually very brief assessments composed of just a few questions, because interview time is at a premium. One example of this in the United States is the Health and Retirement Study funded by the National Institute on Aging, which has a goal of understanding and monitoring the impact of retirement on health and well-being. It is a large-scale, prospective survey of individuals over age 50 and has included several questions to evaluate SWB. The Behavioral Risk Factor Surveillance System, which is a very large cross-sectional telephone survey designed for investigating behavioral risk factors, includes a life-satisfaction question that has been used by researchers (e.g., Oswald and Wu, 2009). It is government-run (by the U.S. Centers for Disease Control and Prevention) and conducted by individual state health departments. The Survey of Health, Ageing and Retirement in Europe, conducted by a consortium of European investigators, has been used to compare eudaimonic and hedonic ratings with each other and across countries (Vanhoutte et al., 2012). The German Socioeconomic Panel and the British Household Panel Study (recently integrated into the UK Household Longitudinal Study) include brief questions on evaluative well-being of the form "How happy are you at present with your life as a whole?" The Health and Retirement Study has been working on developing survey-friendly versions of short hedonic assessments and piloting them in subsamples of the larger data collection. The panel discusses these efforts further in later parts of the report.

1.4 REPORT AUDIENCE, REPORT STRUCTURE

The audience for this report includes statistical agencies, research funding agencies, policy makers, researchers, and the general public. Most of the recommendations in this report are directed toward U.S. statistical agencies that either are already engaged in collection of self-reported well-being information or may do so in the future. However, the report also presents guidance for a research program that is relevant to science and health funding agencies. Additionally, the panel hopes that the report will prove useful to researchers and others interested in the multidimensional nature of moment-to-moment and reflected well-being—something that is much more nuanced and difficult to measure than can be understood simply by asking people if they are happy.

The remainder of the report is structured as follows: Chapter 2 sketches a brief history of measurements of self-reported well-being and their inclusion in survey development; it also defines more technically the evaluative, experienced, and eudaimonic approaches introduced above. Objectives of this careful definition are to clarify the distinctiveness of experienced (and hedonic) well-being from evaluative well-being (life satisfaction) and to assess the relationships among these different dimensions, including the extent to which each measures something unique. The panel also begins its exploration of the implications of this multidimensionality for policy application.

Chapter 3 delves more deeply into ExWB, identifying in greater detail its dimensions and the alternative techniques for measuring them. The panel assesses the state of research on methods for measuring its many dimensions, positive and negative, as well as related sensations such as pain, anger, arousal, etc., across different reference periods, from the momentary to day-long assessments and reconstructions.

Chapter 4 addresses a series of conceptual and measurement issues ranging from cultural and aging effects to survey ordering, context, and mode effects. In the process of discussing difficult survey issues, various types of self-reported bias are identified, along with other aspects of the science that are not well understood. These points in turn suggest a number of research needs, stated throughout the chapter.

Chapter 5 focuses on the potential of measures of self-reported well-being, and particularly measures of ExWB, to inform policy decisions. It identifies what is known about the predictive capacity of these measurement constructs, which in turn suggests what questions can be informed by the data. The panel evaluates current policy uses of the data and promising directions, and it discusses the value of data on these constructs beyond policy (for example, as a general informing and monitoring tool).

Chapter 6 focuses on data collection strategies. It presents an overall approach that involves leveraging existing datasets and modifying ongoing data collection efforts. The panel notes the important role of smaller-scale studies, the use of nontraditional surveys, and new technologies to address specific questions.

Appendixes to the report provide details on some of the key ExWB questions and modules currently in place, such as those in the ONS Annual Population Survey, the HWB-12 Survey, and the Gallup World Survey. Also included as Appendix B is a separate report produced in mid-project by the panel, which was written to assess and provide guidance specifically on the ATUS.

2

Conceptualizing Experienced (or Hedonic) Well-Being

In its narrowest and most traditionally measured form, experienced well-being (ExWB) is the series of momentary affective states that occur through time. In practice, a number of measurement approaches and objectives coexist. These range from the moment-to-moment assessments of affect to instruments that require reflection by respondents about longer time periods, such as how they felt "yesterday." At the longer intervals, well-being assessments are likely to take on the characteristics of "life evaluation" measures. Experience measures can, in a sense, be viewed as a subspectrum of the overall subjective well-being (SWB) continuum, which at one end involves a point-in-time reference period and is purely hedonic ("How do you feel at this moment?") and at the other end involves evaluation of a comparatively very long reference period ("Taking all things together, how would you evaluate your life?"). The ExWB portion of the continuum ranges from the momentary assessments of affect (the shortest framing period) to global-day assessments or day reconstructions at the longer end.[1] As the reference and recall periods lengthen, a measure takes on more and more characteristics of an evaluative well-being assessment. Specification of the reference period has a strong impact on the results of affect questions and, indeed, on what is being measured.[2]

---

[1] Week-long reference periods have also been used in ExWB assessments, particularly in health contexts (e.g., a respondent may be asked about pain last week).

[2] Another consideration for evaluating the associations among the types of SWB is that there may be a confounding of construct and measurement technique. One feature of how some SWB assessments instruct respondents is to ask them to answer for a particular period of time, say, about the past month, the past week, the past day, or about the current moment.

ExWB is defined by people's emotional states but may also include sensations such as pain or arousal, ruminations, a sense of purpose or meaning, or other factors. Hedonic well-being typically is used in association with the narrower, emotional (or affect) component of ExWB. For this reason, the term "hedonic well-being" is—in this report—replaced with "experienced well-being" to convey this slightly broader construct.

ExWB is a term with a very close connection with the much older and extensive field of mood and emotions. A reasonable argument can be made that the terms *hedonic well-being* and *emotions* are synonymous; and sometimes hedonic well-being is called "emotional well-being" (see, for example, Zou et al., unpublished). The fact that they incorporate similar partitions of positive and negative aspects further confirms their similarities. Emotions can be fleeting states that vary from minute to minute; however, when emotions are aggregated over longer periods of time, they become more stable and reliable measures that may better fulfill the needs of well-being researchers. Historically, the "standard" period studied for hedonic well-being or ExWB analysis has been a single day. The initial thinking behind this was that 24 hours was a period that provided some stability and could be assessed without too much concern about recall biases; the panel discusses the implications of these assumptions later, along with the alternatives.

2.1 DISTINCTIVENESS OF EXPERIENCED AND EVALUATIVE WELL-BEING

An important consideration for determining the value of ExWB data and statistics—for research, policy, and general information purposes—is its distinctiveness from measures of evaluative well-being. One might expect people with high levels of overall SWB to report, in most cases, relatively high levels for both its evaluative and experienced dimensions.[3] Very high associations of ExWB with evaluative measures would mitigate the case for regularly including both types of measures in data collections. The goal

---

This is known as the *reporting period*. The problem arises when a hedonic construct such as happiness, which can fluctuate throughout a day, is assessed with a long reporting period, say, "over the past week." Long reporting periods are associated with a shift from an immediate recall of emotions during recent experience to respondents' overall perception of their emotion (Robinson and Clore, 2002). Thus, hedonic SWB measures that use longer reporting periods can start to look more like evaluative well-being measures, creating a confounding effect.

[3] A fairly extensive literature exists on the relationship between evaluative well-being and ExWB. As just one example, Zou et al. (unpublished) found life satisfaction and emotional well-being (their ExWB construct) distinct, though with significant overlap when assessed by multiple indicators.

of this section is to go beyond an intuitive impression of the associations among types of SWB by examining the empirical evidence.

ExWB measures are designed to capture emotions as they fluctuate from moment to moment and in response to day-to-day events and activities. They therefore aim to be reactive to a respondent's immediate focus. For example, for individuals at work, their reported affect is likely related to the immediate task at hand and not to broader issues such as the state of their marriage or their financial circumstances—topics that typically fall more squarely into the evaluative well-being domain. Issues that are only infrequently on a respondent's mind at any particular time during the course of the day (politics, the state of the economy, etc.) are more likely to surface as a measurable effect on SWB upon reflection—as in evaluative measures—or if the respondent is explicitly prompted to consider them. This suggests a significant difference in what is likely to be captured by—and in turn, what is the purpose of—measures of life satisfaction (reconstructed) versus experienced (momentary) well-being. One example of how this difference plays out occurs in measures that track the day-to-day experiences of the unemployed but do not track the unemployment rate.

Just as evaluative well-being and ExWB are conceptually distinct, at the empirical level positive and negative experiences are also separable and influenced by different factors.[4] As detailed below, evidence of this distinctiveness rests not only on correlations and factor analysis but also on multimethod assessments employing measures of SWB beyond self-report surveys. Furthermore, when variables that predict evaluative well-being, positive experience, and negative experience are compared, there are differences in which variables most strongly predict each of these aspects of SWB.

The literature consistently indicates that one aspect of SWB can be moved by a manipulation while another aspect of SWB moves much less or sometimes not at all. Longitudinal research (e.g., Lucas et al., 2003; Stutzer and Frey, 2004) indicates that people experiencing important life events such as marriage or childbirth may react more strongly as measured by one construct compared to another, and over time the different measures show differential patterns of adaptation. Bradburn (1969) found that positive and negative affect are not opposite ends of one dimension but are largely independent of one another; a person can be high on one and either high or low on the other. Bradburn's findings have been replicated many times; for example, Gere and Schimmack (2011) found that, even after controlling for measurement error and bias, positive and negative feelings were distinct. Andrews and Withey (1976) confirmed that life satisfaction is separable

---

[4] Just as, when assessing the economy, more than just gross domestic product is needed to capture its important aspects (growth rates, inflation, employment rates, Gini index, and so on), more than one measure is needed to capture the most important features of SWB.

from positive and negative affect. Lucas et al. (1996), using multimethod measurement (both self-reports and peer reports) and measures over time (a 2-year period), found that evaluative well-being and ExWB were distinct, as were measures of negative and positive experience. Kapteyn et al. (2013), using a specially designed experimental module for the RAND American Life Panel that included measures of evaluative well-being and ExWB, also found life satisfaction and the positive and negative dimensions of ExWB to be distinct, although they found additional factors when different response scales were employed.

Evaluative well-being and ExWB have different causes and correlates as well. Luhmann et al. (2011) found that people react to certain events, such as marriage and childbirth, more strongly in their evaluations of longer-term well-being than in their reports of experienced reactions. However, other events (bereavement, reemployment, and retirement) produced stronger experienced reactions. In examining adaptation to these same events in a meta-analysis of longitudinal studies, the authors found that people adapt more quickly to marriage and childbirth along the ExWB dimension, relative to evaluative well-being, but more slowly to unemployment and reemployment. They found that, for virtually every life event they studied, there was a different pattern for ExWB versus evaluative well-being. For some variables, such as childbirth, they found that ExWB and evaluative well-being could move in opposite directions (Luhmann et al., 2011).

In a review of the evidence, Schimmack (2008) concluded that—even after taking into account measurement error and other factors—life satisfaction, positive experience, and negative experience are to some degree distinct. Thus, people's SWB cannot be fully or accurately captured without assessing all three.

CONCLUSION 2.1: Although life evaluation, positive experience, and negative experience are not completely separable—they correlate to some extent—there is strong evidence that multiple dimensions of SWB coexist. ExWB is distinctive enough from overall life evaluation to warrant pursuing it as a separate element in surveys; their level of independence demands that they be assessed as distinct dimensions.

As discussed in detail in the next chapter, ExWB can and often is usefully parsed into even narrower groupings. For instance, negative feelings can be divided into anger, sadness and depression, and worry and anxiety. Although there is often a tendency to experience these emotions together, and the same people who frequently experience one of them are also likely to feel the others frequently, these different types of negative feelings can be separated. It may be desirable to measure them separately because they are at times associated with different circumstances. However, time limitations in surveys may prevent thoroughly assessing each type of affect, or each subcomponent of the various types of SWB. Furthermore, feelings, such as anger and worry, can be parsed even more finely; the question of how fine the measures need to be is to some extent a practical issue depending on time constraints in administering the survey. However, for some policies, particular forms of ExWB, such as depression or anger, might be particularly salient and should be assessed. What must not be lost sight of is the fact that the dimensions of SWB described above have been broadly studied and much is understood about their structure and causes. Substantial evidence indicates that, in a full assessment of SWB, evaluative well-being (or life satisfaction) and both the negative emotion and positive emotion components of ExWB should be measured separately. If survey time allows, finer distinctions can be assessed within each of these constructs, as outlined below.

What unique information, then, do ExWB measures add beyond that which can be gleaned from evaluative well-being surveys, as well as other economic or demographic measures? It would make little sense to measure ExWB (and, in turn, recommend data collection on it to statistical offices) if it did not add important new information, given that evaluative well-being can be self-reported using one question easily attachable to existing surveys. The panel's position, developed above, is that both the stand-alone content of the ExWB metrics and the information that stems from contrasts between them and evaluative and eudaimonic metrics are potentially valuable for statistical purposes and relevant to a range of policy questions.

Evaluative well-being and ExWB may have different implications for policy (Diener, 2011; Graham, 2011; Kahneman et al., 2006). The latter is more directly related to the environment and context of people's lives. Using data from the Gallup World Poll, Deaton (2012) found, for example, that health state correlates more strongly with ExWB (though it is also important for evaluative well-being) and that marital status and social time are more strongly correlated with ExWB.[5] Other aspects of daily behavior, such as the nature of a person's commute to work and the nature of a person's social networks, are reflected in positive and negative affective states (separable aspects of ExWB). The quality of people's daily experiences is also linked to health status and other outcomes via channels such as worry and stress on the one hand and pleasure and enjoyment on the other.

Evaluative well-being, while also sometimes influenced by these factors, is more likely to reflect people's longer-term outlook about their lives as a whole. It may also be related to, and reflected in, longer-term behaviors such as investments in health and education. The *World Happiness Report* (Helliwell et al., 2012), which focuses primarily on life-evaluation mea-

---

[5] Bradburn (1968) and Bradburn and Orden (1969) also confirm this in their studies of the dimensions of marriage happiness.

sures, documents their closer linkages (relative to ExWB) to life circumstances, which may give them relevance to long-term macro policy making (and perhaps other areas, such as informing service delivery). With respect to the role that children play in most peoples' lives, the differing assessments that come from time-use ExWB metrics (largely negative) versus life evaluations (largely positive) are a good example of how the former capture effects of the day-to-day environment while the latter capture respondents' aspirations about their lives as a whole.[6]

Kahneman and Deaton (2010) found that, in the United States, income correlates more closely with evaluative well-being than with ExWB (they specify "emotional" well-being). The correlation between ExWB and annual income tapers off at roughly $75,000, or roughly the median U.S. income, while the relationship between income and evaluative well-being continues in a linear fashion. After a certain point more income does not seem to make people enjoy their daily lives more (although very low income is clearly linked with suffering and negative moods), but higher levels of income offer people many more choices about how to live and what to do with their lives.

Diener et al. (2010) found that income better predicts life evaluation scores, whereas "psychosocial wealth," which includes factors such as social support and learning new things, better predicts life satisfaction. Their study of Gallup World Poll data showed that income influences life satisfaction but less so than does experience (affect). Positive feelings, such as enjoying life, were more strongly predicted by psychosocial wealth. Similarly, Graham and Lora (2009) found that the most important variables for the reported life satisfaction of the "poor" (respondents below median income) in Latin America, after having enough food to eat, were having friends and family to rely on in times of need. In contrast, the most important variables for the life satisfaction of the "rich" (respondents above the median) were work and health. It is likely that friends and family are the vital safety nets that make daily life tolerable for the poor, while work and health are what provide respondents with more means to make choices in their lives.

Individuals who focus primarily on daily experiences—due to low expectations, lack of agency, or imposed social norms—may have less incentive to invest in the future. In rapidly growing developing economies, Graham and Pettinato (2002) found lower levels of reported evaluative well-being among respondents with relatively high levels of income mobility compared to very poor rural respondents. It seems that people are better able to adapt to unpleasant certainty and retain relatively high levels of evaluative well-being (and likely higher in ExWB than in evaluative well-being) than to live with uncertainty, even when that uncertainty is associated with progress (Graham, 2008, 2011; Graham et al., 2011).

---

[6] See Clark and Senik (in press); see also Dolan (2012) and Graham (2011).

Individuals who have a longer-term focus and are more achievement oriented, meanwhile, may at times sacrifice daily experiences for longer-term objectives and anticipated evaluative well-being in the future. An example is those who choose to migrate to another country to provide their children with opportunities or to participate in social unrest for a broader societal objective. Graham and Markowitz (2011), based on data from Latin America, found lower levels of evaluative well-being among individuals who planned to migrate in the next year—a relatively extreme behavioral choice with future benefit in mind.

Health also correlates differently with different aspects of SWB. Positive affect has been found to predict response to illness (Cohen et al., 2003), with higher levels correlated with lower incidence of cardiovascular disease (Boehm and Kubzansky, 2012). Daily stress and other dimensions of negative affect are positively correlated with illness and with lack of access to health insurance. In contrast, the relation between evaluative well-being and cardiovascular disease, if it exists, is less well known. While ExWB is clearly associated with a reduced likelihood of smoking, the relationship between evaluative well-being and smoking, while still negative, is less consistent (Kahneman and Deaton, 2010).

CONCLUSION 2.2: To a larger degree, evaluative well-being, positive experience, and negative experience have different correlates and (presumably) causes, and can reflect different aspects of life that are relevant to policy. Thus, measuring all in national and specialized surveys is recommended.

To summarize the preceding discussion, sometimes people make sacrifices that lower their ExWB in order to achieve long-run higher evaluative well-being. Conversely, some people may seek greater levels of ExWB and forgo long-run evaluative well-being. This should not be a surprise, as achieving certain overarching objectives, such as advancing a science, completing a doctorate, or performing risky surgery, all of which are likely linked to higher levels of evaluative well-being, are likely to entail an increase in stressful or unpleasant days. Assessing these dimensions separately will shed light on how people view these trade-offs. It will also make evident policies that might affect one type of well-being but not another; assessing evaluative well-being and ExWB as distinct constructs will allow consideration of whether one type of SWB is sometimes bought at the expense of another type.

2.2 DIMENSIONS OF ExWB

The bottom-line question of this section is "what dimensions of experience factor significantly into people's SWB and should therefore be prioritized when designing surveys?" This consideration is crucial in practical terms as organizations such as the UK Office for National Statistics (ONS) begin constructing and fielding SWB modules; it is also a major focus of the recently completed OECD *Guidelines* (2013) project. For the ExWB component of SWB, the most obvious analytic decision for survey design is how to allocate questions between negative and positive affect, but there are other (sometimes more specific) emotions and sensations as well. Going forward, statistical agencies will also be asked to consider additional lines of demarcation that may not fit neatly into the positive and negative emotion categories, some of which the panel discusses later in this section.

2.2.1 Negative and Positive Experiences—Selecting Content for Surveys

Empirical SWB research (see Diener et al., 1999; Kahneman, 1999) strongly supports the separation of positive and negative emotional states. The two dimensions also have different correlates in the general population, which may carry policy implications. A number of researchers—Tellegen et al. (1994, cited in Watson and Clark, 1999) and Diener et al. (1995, cited in Watson and Clark, 1999)—have shown very low raw correlations between positive and negative affect, but higher, though still moderate, relationships after controlling for various random and systematic errors.

CONCLUSION 2.3: Both positive and negative emotions must be accounted for in ExWB measurement, as research shows that they do not simply move in an inverse way. For example, an activity may produce both negative and positive feelings in a person, or certain individuals may be predisposed to experience both positives and negatives more strongly. Therefore, assessments of ExWB should include both positive and negative dimensions in order for meaningful inferences to be drawn.

Additionally, indicators of negative emotion are distinct from one other. Evidence suggests that dimensions of negative affect—sadness, worry, stress, anger, frustration, etc.—tend to be more differentiated than those on the positive side, which tend to move more in unison, carries implications for data collection. Research on how different adjective terms cluster (e.g., Kapteyn et al., 2013) show that negative emotion measures generally seem to have lower intercorrelations than do positive ones and may be subject to more variability as a function of specific adjectives used in survey questions. Positive measures appear to be more robust in this sense. Within the clusters, however, the construct measures appear to be fairly robust with regard to the selection of particular adjectives from the cluster.

The multidimensional character of negative emotion suggests a need for more questions about it (relative to positive emotions) on surveys intending to cover the full range of ExWB. The ONS approach, for example, reflects the positive–negative dichotomy, as the two dimensions are separated in the question structure. However, for now, the ONS surveys still lack multiple dimensions/questions on the negative side.

> **RECOMMENDATION 2.1:** When more than two ExWB questions can be accommodated on a survey, it is important to include additional ones that differentiate among negative emotions because—relative to the positive side—they are more complex and do not track in parallel (as the positive emotion questions tend to).

While "happiness" has received a great deal of attention in the media, and the positive dimensions of SWB are actively researched in the literature on evaluative well-being, a number or researchers have emphasized measurement of negative emotions (suffering),[7] and this choice of focus is to some degree a policy choice. Kahneman's view[8] on the positive–negative balance is that:

> the focus on happiness is misguided and . . . in part is an accident of language. We measure length and not shortness, we measure depth and not shallowness, and we only see in dimensions that are marked on the one side we are thinking of. We should be measuring suffering. And we should act as a society to reduce suffering. . . . I am much less concerned about happiness and [in favor of] reducing human suffering.

At this point, there is not enough empirical evidence about these two dimensions of ExWB to know which is more policy responsive. Clearly, as Kahneman argues, reducing negative experience, particularly prolonged

---

[7] A major exception to the "focus on happiness" is found in the field of mental health, where far more attention has been directed to the negative side of the emotional balance than to the positive side. Western cultural biases, embodied in the psychiatric conceptions of mental health, have led to a concentration on efforts aimed at reducing negative affect and suffering. However, Sheldon Cohen's research (available: http://www.psy.cmu.edu/~scohen/AmerPsycholpaper.pdf [October 2013]) has examined how positive emotions/attributes lead to higher resistance and identifies specific mechanisms through which different types of social constructs influence physical health (including social stress and immune function issues)—revealing a potentially powerful effect.

[8] From a talk, available: http://stevensonfinancialmarketing.wordpress.com/2013/04/11/8345 [October 2013].

suffering, is often a rightful policy objective, even if the exact policy levers have not been identified. Knowing more about the relationships between determinants and negative experience is important contextual information.[9] If data can reveal the links, it can be left to researchers to discover if policy could be creatively used to have an impact. A targeted policy to assist the poor may focus on negative experience, but it may be linked to positive affect as well.

This line of reasoning suggests the value of framing measurement in terms of experience, which can reasonably include pain and other sensations that factor into suffering but may be omitted by a narrower hedonic approach. In other words, measuring "experience" seems essential for addressing issues of long-term suffering in various populations. Relatedly, the metric for characterizing emotions and suffering—that is, ExWB—could be based on the duration of the day (or other time period) spent in that state. Of course, it is not clear that all methods for capturing ExWB are capable of yielding such temporal metrics; it probably requires momentary assessment or reconstruction of a day to achieve duration-weighted indices.

RECOMMENDATION 2.2 (Research): A scale of suffering that has a duration dimension would be a useful measurement construct and should be developed. Such a measure might capture and distinguish between things like minutes of pain or stress versus ongoing poverty, hunger, etc. Suffering is not the absence of happiness or the presence of only negative experience, and the scale should reflect this in a way that suggests relevant classes of policies. Little work has been done on a scale of suffering, so the research will have to begin at the conceptual level. This research should examine the information content of alternative descriptive adjectives, some of which have perhaps not yet been used in the literature on SWB, but which could round out the set.

---

[9] While not an emotion scale and thus not a measure of SWB (though likely a predictor of it), a scale of negative life events was developed for use in the General Social Survey (GSS) as a component of the GSS index of societal well-being. This approach, by registering exposure to the negative circumstances and events experienced by people (e.g., hospitalization, death of a family member, eviction, crime victimization), was designed to report "objective experiences that disrupt or threaten to disrupt an individual's usual activities, causing a substantial readjustment in that person's behavior" (Thoits, 1983). As described by Smith (2005), this approach has been used extensively not only to account for differing levels of reported well-being among individuals or groups but also for understanding and predicting individual illness (both psychological and physiological); in so doing, it provides "factual data for the formulation of public policies to deal with these problems" (p. 18).

A focus on suffering may also resonate with the public in a way that discussions of happiness do not. When people are asked what is more important for programs and policy, reducing suffering (which requires monitoring negative affect) or increasing happiness (which requires tracking positive affect), the available evidence suggests that the majority prioritize the former. Dolan and Metcalfe (2011) surveyed people to ask whether government policy should seek to (1) improve happiness or (2) reduce misery, and there was more support for the second option. Such findings have important implications for how happiness (and misery) are discussed in the popular press and public debate. Thus, while it is not obvious that a good measure for suffering has yet been developed, it may be politically more acceptable to aim policies at reducing negative affect as opposed to increasing positive emotions. In fact, monitoring (and reduction of) suffering and stress has been the more common objective of government policy.

RECOMMENDATION 2.3: Given the importance of both positive and negative experience, the one-dimensional term "happiness" should not be used to label most ExWB measures. Another limitation of the term is that it is often also used as a descriptor in evaluative measures, which creates another likely source of confusion. Instead, including a term signifying misery or suffering in addition to positive emotions would be more balanced.

For the fullest possible descriptive accuracy, having two words (one for positive and one for negative experiences) incorporated under the ExWB dimension of SWB has an advantage, even though it is well understood by researchers in this field that "hedonic" refers to both positive and negative experiences (again, more broadly defined than emotions). While it may or may not be intuitive that there are both increases and decreases in "well-being," it is clear that SWB measurement is about much more than "happiness." This general point applies to measures of evaluative well-being as well, where an overarching term such as "happiness" can easily mask the great depth of findings in well-being research. It seems clear that labeling (word choice), especially in the popular press, does influence the public debate. Linguistic biases need to be addressed, both in survey construction and in presentation of information. Certainly, labeling measures (or measurement programs) as "well-being/suffering" would dilute the relentless focus on the positive.

There are alternative ways to characterize the positive and negative sides of ExWB—most notably as unidimensional or bidimensional. "Balance" metrics have also been used that combine positive and negative poles. Bidimensional (and possibly multidimensional) approaches offer a more relevant concept—relative to unidimensional measures such as happiness—because these measures give a richer picture of experience and the possible environmental factors that might influence affective experience. For example, people who report suffering directly may be different from those reporting low levels of satisfaction. Insofar as the various dimensions of emotions are driven by different factors, measuring them separately offers more insight into situations where polices potentially could improve SWB. Balance measures also allow those, such as policy makers, who want to concentrate on reducing negative experience to still have access to information underlying that end of the spectrum. There are issues in how to define the balance measure: Is it just an average? Is it enough to know that a respondent had x minutes of positive affect and y minutes of suffering during that day? Or do we need hours, events, or intensities? A counter-argument in favor of the value of unidimensional measures is that they allow people to scale or integrate for themselves how the different measures of emotions or different aspects of their lives should be weighted.

A balance concept could encourage investigating actions that might increase positive emotion (and possibly thereby increase positive aspects of SWB, and possibly health), as well as actions that might reduce suffering. Such a concept need not presume that suffering is the opposite of happiness, because it is possible to be moderately happy while experiencing a moderate degree of suffering as well as a moderate degree of positive emotion. It is not yet clear exactly which balance metrics would be appropriate and therefore should be considered for national statistics.

2.2.2 Eudaimonia

Beyond and possibly intertwined with evaluative well-being and ExWB are additional types of psychological well-being or SWB that may be of potential interest to policy makers, leaders, and citizens. A number of alternative or supplemental forms of psychological well-being have been placed under the rubric of *eudaimonic well-being*: these include optimism; quality of social relationships; meaning and purpose in life; mastery, skills, and achievement; freedom to make decisions regarding one's own life; engagement, interest, and flow; and self-worth. Eudaimonic well-being comes into play if one assumes that people commonly strive for more than just "happiness" and one believes a worthwhile societal goal is to encourage citizens to pursue meaning and purpose in their lives, to give and receive social support, and to have skills and self-esteem.[10]

---

[10] The literature on "noncognitive skills" has addressed the role of some of these individual personality traits and abilities in various outcomes, such as labor market or educational success. Heckman et al. (2006) list as noncognitive skills various social skills, time preferences, motivation, and the ability to work with others.

The Ryff Multidimensional Scales of Well-Being (Ryff and Keyes, 1995) is an example of a widely used, predominantly eudaimonic scale; it consists of six dimensions of wellness (autonomy, environmental mastery, personal growth, positive relations with others, purpose in life, self-acceptance). However, the underlying latent structure and factorial validity of this model remains highly contentious; specifically, there is evidence of a high correlation (lack of distinctiveness) among four of the six dimensions (Springer et al., 2006).[11]

Eudaimonic well-being is broadly related to the opportunities that people perceive they have to exercise choice and to pursue fulfilling lives. While distinct,[12] eudaimonic well-being may also figure into assessments of both evaluative well-being and ExWB. For example, perceived meaning attached to one's job or taking care of one's child may play a role in a person's self-reported well-being, or it may be a factor in predicting whether a person will continue to engage in an activity that scores poorly in a momentary assessment. While eudaimonic well-being is surely important and worth measuring, the field has much less experience with metrics for this type of well-being, and further research and testing are necessary before recommending its inclusion in large-scale surveys.

The evidence for the independence of subtypes (or dimensions) of eudaimonic well-being from each other and from SWB is more limited than it is for evaluative well-being and for the dimensions of ExWB. The new OECD guidelines on measuring SWB include a separate measure of eudaimonic well-being. Literature cited in that volume suggests that eudaimonia correlates less closely with the other SWB measures than do measures of positive or negative affect or of life evaluation. Gallup World Poll data for the OECD countries show the highest correlation between positive and negative affect (–0.39) and the lowest between purpose (the Gallup organization's measure of eudaimonia) and negative affect (–0.09).

---

[11] Diener et al. (unpublished) have recently begun developing a "Comprehensive Psychological Well-being Scale," which includes a eudaimonic (meaning and purpose) component. In reference to the debates about Ryff's factor structure (which does not show six clearly differentiable factors), they are conducting a factor analysis of the new scale to demonstrate the correlation and separability among self-assessment components, which fall into four categories: Relationships (perceived support, social capital, trust, respect, loneliness, and belonging); Mastery/Engagement (flow, engagement, interest; using one's skills; learning new things; control of one's life; and achievement, accomplishment, and goal progress assessments); Meaning and Purpose in Life; and Subjective Well-being (optimism, life satisfaction, positive feelings, and negative feelings).

[12] The distinctiveness of SWB components may be present even at the cellular level. Fredrickson et al. (2013) investigated "molecular mechanisms underlying the prospective health advantages associated with psychological well-being" and found that "hedonic and eudaimonic well-being engage distinct gene regulatory programs despite their similar effects on total well-being and depressive symptoms" (p. 1).

Life satisfaction has a correlation of approximately 0.23 with positive affect and −0.23 with negative affect; its correlation with purpose is 0.13 (OECD, 2013, pp. 33-34).

The purpose (or lack of it) dimension of eudaimonic well-being seems particularly important, as it is associated with much of what we do. Purposefulness (or worthwhileness) can be an important driver of behavior and is experienced in much the same way as emotion. And of course, an ExWB measure might capture some purpose dimensions. In thinking about the full dimensionality of SWB, the concept of "worthwhileness" or "meaningfulness" has been given considerable attention in the literature and has apparently been deemed central to it by ONS, which includes a question on eudaimonic well-being in its SWB module. This dimension may be important for understanding (or predicting) why and when people engage in various activities during the day or in life more generally.

For example, a parent may be less unhappy changing a child's diaper because he finds taking care of his child a worthwhile activity. Or reading the same story over and over to one's children may not always bring a great deal of pleasure, but it is purposeful (or worthwhile, meaningful, fulfilling, rewarding). And the reader (parent) feels that purpose *at the time*. Activity-based data suggest that time spent with children is relatively more rewarding than pleasurable and time spent watching television is relatively more pleasurable than rewarding—but both are drivers of behavior (Kahneman and Krueger, 2006). A rich conception of the flow of feelings places both pleasure and purpose on experiential footings.

Calling purpose a feeling suggests that it is an emotion that can be placed on a comparable footing to more recognized emotions like joy, anxiety, anger, etc. To most psychologists, feelings *are* emotions and so any feelings of purposefulness (or purposelessness) would simply add to (or subtract from) the overall "goodness" of an emotional experience. Feeling that something is purposeful (or purposeless) adds to (or subtracts from) the overall "goodness" of the sentiments associated with an experience. This is somewhat related to Fred Feldman's attitudinal hedonism (Feldman, 2004).

For measurement, it may not make much difference whether one thinks of purpose as contributing directly to good and bad emotions or as sitting alongside but separate from them, as a distinct sentiment. What matters is that the adjectives for purpose (fulfillment, etc.) are distinct from those used for pleasure (fun, etc.) and that a range of good feelings (good emotions, good sentiments) contributes to overall well-being. Hedonic emotions and purpose are both felt experiences and ideally both would be measured. If anything, the purpose dimension is a simpler construct than other emotions in that it is largely nonaroused and so either good (purposeful, worthwhile, meaningful, fulfilling) or bad (pointless, worthless, meaningless, unfulfilling). One would not need to measure both if pleasurable experiences, for example, were highly correlated with purposeful ones. The evidence on this issue is scarce, but data from the Day Reconstruction Method suggest that, while some activities are high in both purpose and pleasure (e.g., exercising) or low in both (e.g., commuting), others are high in pleasure and low in purpose, such as watching television, and others are low in pleasure and high in purpose, such as volunteering for unpleasant tasks (White and Dolan, 2009).

> CONCLUSION 2.4: An important part of people's experiences may be overlooked if concepts associated with purpose and purposelessness are not included alongside hedonic ones like pleasure and pain in measures of ExWB. Crucially, central drivers of behavior may also go missing. People do many things because they are deemed purposeful or worthwhile, even if they are not especially pleasurable (e.g., reading the same story over and over again to a child, visiting a sick friend, volunteering); they also do many things that are pleasant even if they are not viewed as having much long-term meaning in the imagined future.

In terms of relevance to policy, there appear to be differences in the way ExWB and evaluative well-being relate to time spent volunteering when purpose is accounted for. Greenfield and Marks (2004) found that among an older population group, volunteering was associated with more positive ExWB but not with significantly less negative affect. The extent to which volunteering makes people happier is unclear, as is the extent to which happier people tend to engage in more volunteerism. However, the latter seems to be part of the story, as the measurable association is reduced considerably when fixed effects are controlled (Meier and Stutzer, 2006). This has potentially important implications for those in government trying to "sell" the idea of helping others; it may have more traction if it is presented as a way of increasing happiness (or decreasing unhappiness) through purpose. In any case, more can be said about how well life is going if purpose is accounted for, as well as pleasure.

Not accounting for purpose alongside pleasure is potentially a threat to the legitimacy of ExWB measures. One of the attractions to policy makers of the constructs of evaluative and eudaimonic well-being is that they allow for consideration of nonhedonic sources of happiness and suffering. But beyond the reflective exercise of asking respondents to consider their lives overall, additional insights can be gained by assessing degree of purpose, or lack thereof, as it is revealed during such life experiences as looking after friends or family (or not having those connections), pursuing goals (or not having them), or dedicating oneself to work (or finding work pointless).

RECOMMENDATION 2.4: Where possible, adjectives of purpose can—and should—be added to experiential assessment methods and techniques, although more needs to be learned about them. Cognitive testing and other psychometric work is needed to find out what members of the public make of these and other possible descriptions. Also needed is quantitative analysis of the correlations between candidate descriptions.

2.2.3 Other Candidate Emotions and Sensations for Measures of ExWB

In thinking about exactly which adjectives best capture the ExWB constructs of interest and that warrant measurement, it is important to consider those that may not sort neatly as positive or negative emotions. Again, thinking in terms of experiences and not just emotions allows for inclusion of more of these factors. Whether or not an ExWB measure should include factors beyond the realm of emotions depends on the research or policy question at hand. For example, sensations such as physical pain, numbness, heat, or cold could be part of the conceptualization of ExWB at the momentary level of measurement—particularly if the context is people's health or housing conditions. Certainly, people experiencing pain will on average report higher levels of negative well-being, all else being equal (Krueger and Stone, 2008). The 2012 Health and Retirement Study is a good example of a survey module that asks about negative emotions and physical pain, as is the 2010 version of the American Time Use Survey. The National Health Interview Survey and the National Health and Nutrition Examination Survey, major data collection programs of the National Center for Health Statistics of the Centers for Disease Control and Prevention, are other good candidates.

RECOMMENDATION 2.5: Pain may be an important dimension of ExWB given that it affects people's ability to engage in day-to-day activities. Therefore, while still experimental at this stage of research, pain questions should be included in ExWB questionnaires, particularly in domains such as health or housing where this information is particularly germane to research and policy questions.

Ultimately, for a given question, how one characterizes "the momentary" will dictate the value of additional experience considerations such as feelings of pain, spiritual elevation, flow, love, etc. The Positive Affect Negative Affect Scale, often called "PANASX," is a popular emotion scale for which several other adjectives have been identified and tested, such as those related to hostility, guilt, fear, joviality, serenity, shyness, self-assurance, fatigue, surprise, and attentiveness. Anger is particularly complex; Bradburn (1969) and Harmon-Jones (2004) found that anger appears to be largely unrelated to global measures of either positive or negative affect. In principle, surveys can ask about the presence of these states at the momentary level of analyses, but this has not, in the panel's experience, usually been done.

Another candidate for consideration is a crosscutting dimension of affect known as "activation-deactivation" (Larson and Diener, 1992). There is considerable evidence that the range of emotions can be usefully characterized as a two-dimensional space, with high and low arousal as one of the dimensions and positive and negative emotion as the second (see the "circumplex" model of affect, Watson et al., 1988). Arousal is especially relevant when measuring affect in populations that are ethnically and/or age diverse. We discuss this in more detail below, in section 4.1 on cultural effects.

While there may not be enough evidence to include questions about sensations and other factors (beyond emotions) influencing people's experiences in broad surveys, in the same way that questions on evaluative well-being or "feelings yesterday" have been added, there are cases where such factors are clearly relevant and should be included; for example, in assessing pain and mobility in surveys of the elderly (Health and Retirement Study) or in measuring arousal in cross-country comparisons. Descriptors for pain and anger are among the most prominent adjectives of interest beyond the "hedonic" that may not always fit into a positive/negative emotion construct. Other factors, not discussed here and not well understood, may also influence people's moment-to-moment experiences; Box 2-1 describes one example: what respondents happened to be thinking about, which may take the form of intrusive or fleeting thoughts.

BOX 2-1
Pop-Ups: The Likely Importance of Intrusive Thoughts

One of the main benefits of assessments of ExWB is that they overcome some of the focusing-effect problems associated with global assessments of life satisfaction. The focusing effect is used to explain the higher effect that income has on life satisfaction as compared to the Day Reconstruction Method (DRM): Kahneman et al. (2006, p. 1908) conclude that *"the belief that high income is associated with good mood is widespread but illusory."*

Yet the DRM's attempt to capture experienced utility may create a focusing effect of its own by asking respondents to report their feelings when thinking about the *activities* in their lives. This formulation neglects the way in which people's attention drifts between current activities and concerns about other things. It therefore may be useful to consider and measure the impact on experienced utility of important *pop-ups*: things that pop into people's heads as they go about their daily activities but which are not captured especially well by routine assessments of affect.

One of the great advantages of ExWB measures is that they seek to capture the flow of experienced utility over time. Experienced utility is largely influenced by where attention is directed, sometimes voluntarily (such as when an author is focusing on writing a sentence) and at other times involuntarily (such as when the author's children just popped into his head). Measures of emotional well-being do a good job of picking up feelings in general but often miss important *pop-ups.*

Just like purposefulness, these *pop-ups* (also called intrusive thoughts or mind-wanderings) potentially drive a lot of behavior. People can be expected to give a lot of weight to those things that grab their attention, even if they do so only fleetingly. Most of the research to date has been conducted on clinical populations, but there is some evidence that general mind wanderings are frequent, occurring in up to 30 percent of randomly sampled moments during an average day (Smallwood and Schooler, 2006). Generally negative intrusive thoughts have a negative association with well-being (Watkins, 2008). The relationship is not straightforward, however, because the suppression of intrusive thoughts can make things even worse (Borton et al., 2005). There is also some suggestion that even unwanted thoughts may still play an important adaptive role in problem solving and learning (Baars, 2010). On the other hand, it has been suggested that even positive intrusive thoughts can be a source of unhappiness; our ability to think about something other than what we are doing at the present moment comes at an emotional cost (Killingsworth and Gilbert, 2010).

The presence of intrusive thoughts about health helps to explain the difference between experienced utility and decision utility in the valuation of health states (Dolan, 2011). In particular, *pop-ups* about health could explain why people are often willing to make large sacrifices in life expectancy in order to alleviate conditions for which there is a considerable degree of hedonic adaptation (Smith et al., 2006). Recent research has also shown how intrusive thoughts can be used to change behavior. By using therapies that focus on shifting attention, researchers were able to reduce intrusive thoughts about smoking and see positive results on patients' ability to stop smoking (May et al., 2011).

Policy makers will be interested in the consequences, as well as the causes, of intrusive thoughts. In a health context, for example, negative thoughts such as worry are associated with increased cortisol levels and increased heart rate, and they may actually cause increased heart disease and fatigue and slower recovery from surgery (Watkins, 2008).

Of course, negative thoughts might also be good for health through their effects on health-promoting behaviors. There is evidence, for example, that increased worry about breast cancer is associated with a greater probability of undertaking screening (Hay et al., 2006). Among other things, new data will enable policy makers to better determine when to reduce negative thoughts, by how much, and for whom.

One of the attractions to policy makers of "happiness" as represented by evaluative SWB measures is that it allows people to consider the importance of a range of things, including intrusive thoughts. But it does so in rather artificial and abstract ways by asking respondents to consider their life overall. It is much better to pick up the effects of *pop-ups* where they really show up—in the experiences of life.

Every survey will focus attention in one way or another. It seems unlikely that *pop-ups* can be picked up effectively by asking people to focus attention on them. To be more specific, a future DRM-type study could certainly ask respondents about thoughts and feelings before asking what the respondent is doing about them and who they are with. It may be that asking people about their main activity before asking them about their mood draws their attention away from what they were thinking about. Such a study would allow researchers to explore the importance of focusing effects as they relate to current activity and to general mood.

3

Measuring Experienced Well-Being

A number of techniques have been developed and used for obtaining experienced well-being (ExWB) data from subjects. These include momentary assessments that take place throughout the day, such as the Experience Sampling Method (ESM) and Ecological Momentary Assessment (EMA); overall assessments of a day—which may follow an end-of-day or a yesterday structure; and reconstructions of a previous day's activities followed by well-being assessment for each episode, such as the Day Reconstruction Method (DRM). These approaches vary in depth of information and precision of measurement; they also vary in terms of respondent burden. In this section, we review the major techniques used to measure people's ExWB. Some attempt to capture emotional states in the moment; others rely on longer reference or recall periods and thus require some reconstruction or reflective assessment by respondents, pushing them along the time frame continuum toward life evaluations.

3.1 ECOLOGICAL MOMENTARY ASSESSMENT

ESM is a research methodology that asks participants to stop at certain times and make notes of their experience in real time—it measures immediate experience or feelings. EMA refers to a class of methods designed to track emotions associated with experiences as they occur, in everyday life; they thus avoid both reliance on memory and context effects caused by artificial environments (e.g., a laboratory). EMA also provides a methodological framework for capturing other data in the field, such as the use of ambulatory monitors of various physiological states.[1]

In the most paradigmatic type of EMA, respondents provide subjective assessments of their emotions and experiences in real time, as they go about their daily lives (see Shiffman and Stone, 1998, for a review). The usual method is for the respondent to wear an electronic device for a period of time, such as a week, that prompts the wearer at various times throughout the day to respond to a brief survey.[2] Answers are input directly into the device. From EMA data, researchers are able to compute average levels of variables of interest (thus avoiding the problem of relying on participants to aggregate their own experiences) and can also explore peak and diurnal experiences. A primary advantage of this method is that it does not rely on memories constructed after the fact. EMA methods provide direct, subjective assessments of actual experiences, allowing the various biasing factors associated with recall to be bypassed.

On the downside, the intensive nature of EMA studies makes it very difficult to scale up this method to the level of nationally representative surveys. Devices must be provided to (and usually returned by) respondents, who must be trained in their use. Given the considerable respondent burden involved, response rates may be low—especially among some vulnerable or distressed groups—and participant compensation costs are likely to be substantial. Response rates may be especially low among people who have the most difficulty using the devices (e.g., those with

---

[1] For this report, the panel uses the term "EMA" to refer to the class of methods that includes both EMA and ESM.

[2] In its original form, EMA uses randomly selected intervals to avoid bias that could be incurred by a fixed interval schedule. In addition, the traditional EMA approach asks respondents to assess their experiences *right now*, to avoid reliance on memory. However, this approach means that distinctive but important events may not be captured. This problem is exacerbated by the fact that, while the prompts are random, nonresponses to the prompts—missing data—may not be. Respondents may not wish to respond to the device if it beeps at particularly positive (e.g., watching one's child's first steps) or negative (e.g., a fight with one's spouse) moments—yet these are precisely among the moments that researchers would probably want to capture.

One answer to the problem of "missed" events is to alter the EMA protocol from a "right now" sampling frame to a *coverage frame*. In a coverage frame, respondents are asked to characterize their experiences "since the last prompt." Often in this case, a fixed sampling interval is selected, to standardize the length of time respondents are being asked to summarize. The drawback here is apparent; once again, participants are being asked to remember and summarize—albeit over a much shorter time frame (typically just a few hours). But the potential advantage is also apparent: transient but important experiences are more likely to be captured. Coverage sampling frames are probably most useful for smaller samples and when a researcher wishes to capture experiences that are known to fluctuate within the day but may be somewhat infrequent (pain in some patient populations, suicidal thoughts, etc.).

certain disabilities), although technological innovation may make this less of an issue over time.[3]

CONCLUSION 3.1: Momentary assessment methods are often regarded as the gold standard for capturing experiential states. However, these methods have not typically been practical for general population surveys because they involve highly intensive methods, are difficult to scale up to the level of nationally representative surveys, and involve considerable respondent burden, which can lead to low response rates. For these reasons, while momentary assessment methods have proven important in research, they have not typically been in the purview of federal statistical agencies.

The panel notes that this conclusion reflects the current (or past) state of monitoring and survey technology, which is of course changing rapidly. Thus, we would append some important qualifying statements to Conclusion 3.1:

- The ways in which government agencies administer surveys are surely going to change, and as monitoring technologies continue to evolve rapidly, new measurement opportunities will arise. Considered in terms of comparative respondent burden, it may become less intrusive to respond to a modern electronic EMA device (or smartphone beep) than to fill out a long-form survey. So, while EMA may not be practical for the American Community Survey or Current Population Survey for the foreseeable future, real-time analyses may become practical for a number of other surveys, particularly in the health realm. One way this works in practice is that, at various (usually irregular) intervals, respondents would be beeped. The National Institutes of Health is interested in this kind of technology for real-time health applications. As technology advances, such modes could become feasible, even for large-scale surveys at reasonable cost.
- Large-scale (more general) surveys could build in the possibility of mapping the data from single-day measures with the data from more detailed studies for a subset of the sample.
- Experiences in real time, because they are especially relevant to health, have been incorporated into health examination surveys,

---

[3] Differential response rates to questionnaires by subpopulations are a concern generally, but particularly when making group comparisons. Moreover, it is not just an issue for how to interpret measures once they are collected; these group differences may also need to inform the construction of the survey instruments themselves. Some peripheral insights can be learned from Abraham et al. (2009), who showed that high variability exists for survey estimates of volunteering due to the "greater propensity of those who do volunteer work to respond to surveys."

so there is precedent. It is also possible to monitor blood pressure and other physical signals related to affect in real time.

Again, the appropriate methodology and data collection instruments will be framed by what questions are being asked and which policies are to be informed.

3.2 SINGLE-DAY MEASURES

To date, the most common method of measuring ExWB in large-scale survey research is based on assessments of a *single day*. Although single-day measures are currently the standard in survey research, there are also alternative methods, as discussed in this section and elsewhere in this report. A potential criticism of the method is that there is considerable day-to-day variation in hedonic states (people have "good" and "bad" days) and, thus, a single-day assessment might be "too variable." In this section, the adequacy of single-day ExWB measures for testing different hypotheses (e.g., between-group differences) is evaluated. The importance of this topic is that, if single-day ExWB measures are found to be credible for research, the case for including them in large-scale national surveys is strengthened. However, if single-day measures do not approximate ExWB as captured in the more intensive momentary approaches, then, depending on the survey objectives, the case for including them is undermined.

A number of national and international surveys have used single-day assessments to measure ExWB—that is, assessments that target affect for a single day. For example, in the United States, the Health and Retirement Study, the Disability and Use of Time supplement to the Panel Study of Income Dynamics, and the Gallup-Healthways survey employ single-day hedonic assessments, as do the English Longitudinal Survey of Ageing and the surveys on well-being of the UK Office for National Statistics (ONS).

3.2.1 End-of-Day Measures

End-of-day subjective well-being (SWB) measurement is a well-established and frequently used research method. The objective of end-of-day measures is to capture a respondent's assessment of affect for an entire day, which is quite different from the goal of the momentary assessments, although the target objective for methods such as EMA is also often to add up to a full-day measure. ExWB measures are somewhat sensitive to what people are doing at the time of questioning (see Schneider et al., 2011). Compared with momentary assessment, an end-of-day method shifts measurement from a temporal integral of experienced affect to the respondents' summary impressions of how good their day was. There may be some questions and policies for which a reflective assessment of this sort is the relevant measurement objective.

End-of-day self-reports of the form "Overall, how happy would you say your day was?" can be influenced by the same variables that drive answers to evaluative well-being questions, though presumably to a much lesser degree; and they will reflect both the respondent's mood at the time of judgment and the most memorable moments of the day. One way to establish credibility for an overall day measure is to see how well it approximates an "integral" over 1 entire day of momentary assessments. This kind of credibility for single-day ExWB measures would seem to be a prerequisite for including them in large-scale national surveys.

Although research generally indicates that end-of-day measurement methods, typically used in smaller-scale studies, yield credible and consistent data about people's experiences for the day, important research questions remain. For example, more needs to be known about how momentary or activity-based experiences map into longer-period assessments and about how different time increments are remembered by respondents. The usual issues—from salience and recency to duration neglect—apply as well. Some evidence of how respondents weight the day's moments comes from the observation that patients' end-of-day ratings of pain are more influenced by the last EMA measurement of the day (Schneider et al., 2011). This finding is consistent with many other studies (e.g., Redelmeier and Kahneman, 1996; Stone et al., 2000) that show higher recalled pain during episodes of higher concurrent pain. A day that ends well will almost certainly be reported as a better day than a day that ends poorly, even when the averages are identical.

One limitation of end-of-day measures (and a reason that they have not been used more by statistical agencies) is that large population surveys often depend on telephone interviews conducted throughout the day, not just at the end of the day. Because of the survey timing requirement, end-of-day instruments have typically been less practical for use in general surveys than global-yesterday methods (discussed next) have been. However, newer technologies, such as use of interactive cellphone assessments, may offer solutions to some of the data collection constraints associated with end-of-day methods.[4] Mode of survey administration is a central issue when assessing the appropriateness of SWB measurement approaches that require precise timing, as do end-of-day measures.

---

[4] The panel returns to the potential role of new technologies in SWB survey methods in section 6.3.

3.2.2 Global-Yesterday Measures

Global-yesterday measures ask respondents about emotions and feelings experienced the previous day. The concerns about end-of-day measures, in terms of approximating a time integral of real-time emotions, apply here as well, except that the increased temporal distance may accentuate them. That is, relative to global-yesterday assessments, one would expect end-of-day measures to correlate more closely with ESM. However, there has been little systematic experimentation into how the recall and contextual influences act differentially between end-of-day and global-yesterday measures.

As noted above, momentary assessment data collection has typically not been feasible for nationally representative government surveys because it involves considerable respondent burden, which can lead to low response rates. Similarly, end-of-day instruments (usually defined as "before bed") have the practical disadvantage that the survey must take place somewhat precisely at the end of the day. Thus, global-yesterday measures are often the default approach for large surveys. In part because an interviewer can call at any point during the (following) day, global-yesterday questions have featured regularly in large surveys such as those conducted by the Gallup Organization and, more recently, by the ONS.

Christodoulou et al. (2013) validated a global-yesterday version of an ExWB measure by comparing the results against the same emotion adjectives administered using a DRM that links assessments to specific episodes of the day (this work is described in more detail later in this section). Although not the same as testing a single-day assessment against momentary assessments, the idea was that the DRM reconstruction techniques and the duration-weighted average of reported emotions over the previous day should be closer to the "truth" than the global-yesterday measure, which is usually completed quickly and without much contemplation of the day's events. This study shows promise that the correspondence between global-yesterday measures and DRM is good—in this case (depending upon the adjective) it was in the 0.7 range.

Much of the evidence for the utility of global-yesterday measures has been established through research using the Gallup Organization's datasets. These data have generated insights into which groups of the population report being happier from day to day (e.g., middle-aged versus young or old, married versus unmarried, employed and unemployed) or at what times (e.g., weekends versus weekdays, holidays versus work days). Using the "yesterday" measures from Gallup surveys, Kahneman and Deaton (2010) found that income was related to ExWB in nonlinear ways. Stone et al. (2010) found that various measures of ExWB were related to respondent age in patterns that were very different from a measure of life evaluation (evaluative well-being). Deaton (2012) found that the negative impact on average self-reports of ExWB at the time of the 2008 financial collapse in the United States was short-lived. Stone et al. (2012) used global-yesterday measures to extend knowledge about day-of-the-week associations with ExWB measures. Using the Gallup data from 2008 for more than 340,000 U.S. citizens, the authors found contrasts in mood between weekend days and weekdays, but no significant difference in mood on Mondays compared with Tuesdays, Wednesdays, and Thursdays. Some of the effects contrasted quite sharply; for example, 60 percent of the individuals in one age group reported being stressed for much of the day while the figure was only 20 percent in another age group (Stone et al., 2012).

> CONCLUSION 3.2: Global-yesterday measures represent a practical methodology for use in large population surveys. Data from such surveys have yielded important insights—for example, about the relationships between ExWB and income, age, health status, employment status, and other social and demographic characteristics. Research using these data has also revealed how these relationships differ from those associated with measures of evaluative well-being. Even so, there is much still to be learned about single-day measures, and it is possible that much of what has been concluded so far may end up being contested.

These positives notwithstanding, data from global-yesterday, and single-day methods more generally, provide less information about why these differentials exist or during which activities people are suffering more or less. Global-yesterday measures are therefore limited in terms of creating a more detailed understanding of the drivers of ExWB over the course of the day (e.g., variation at the individual level). For this level of analysis, time-use or activities-based data—for example, data generated by DRM-type methods, discussed in section 3.3.1—are needed. Describing group variation, which global-yesterday measures have been shown to do well, is different from explaining the sources of differences in level or the drivers of change for a population.

3.2.3 Appropriateness and Reliability of Single-Day Assessments of ExWB

Survey purpose will dictate the appropriateness of single-day methods (SDMs). Findings that end-of-day ratings correlate well with averages for the day collected using EMA (e.g., Broderick et al., 2009, for pain and fatigue) do not imply that decontextualized end-of-day ratings are sufficient for *all* purposes. For policy, it is often essential to know what experiences or activities are driving affect or changes in affect, and how. That said, low-burden ExWB ratings from a single day may capture some activities and life events well. For example, it may be possible to meaningfully measure emotional effects associated with Valentine's Day (Kahneman and Deaton, 2010) using daily affect questions because that is in a real sense the relevant reference unit of time. And some phenomena that unfold over long periods, such as the cumulative emotional impact of being unemployed or of marital or financial problems, may be captured more in responses to questions that involve reflection than they would be in momentary assessment.

Cross-sectional, single-day surveys are most often used to address group differences in ExWB—for example, are older people happier than younger people? Are females more stressed than males? Or, do males report more happiness than females? The main selling point of single-day measures, upon which the case for their inclusion in large surveys hinges, is their ability to accurately detect group differences in a minimally burdensome way. For this reason, the panel considered the analytic value of SDMs in the context of making between-group comparisons. Examining the contributors to SDM variability helps to address this consideration.

Several sources contribute to the variability in SDMs. If a SDM is assumed to contain a portion of its score that is related to the group factor being explored (e.g., gender differences), then this part of the score presumably remains stable from day to day. In other words, if males are in truth happier than females, that information is embedded in SDM values. However, there are many other factors that impact a particular daily score, including the various events that occur on the day the measurement was taken. The daily variability of a SDM, presumably driven by daily occurrences, can make detection of a group effect more difficult.

One way to measure the level of daily variability is to compute the ratio of over-days variation (that is, with SDMs repeated daily for some respondents) to all variation (that is, the total variation due to daily variation, between-person variation, measurement error, and so on). In one study where several hedonic states were assessed daily for 1 week, Stone et al. (2012) found, using IntraClass correlations, that 30 to 50 percent of all variation was attributable to day-by-day variation. The researchers concluded that most of the day-to-day variation is "real," in the sense that daily events and well-known cyclical effects (e.g., weekday to weekend cycling) were producing it; therefore, it was not reasonable to assume that all, or even most, of the daily variation is measurement error.

CONCLUSION 3.3: Preliminary work suggests that SDMs of measuring ExWB are appropriate for many purposes and contain a valid signal that can be captured by survey studies. Thus, despite their variability, SDMs can be used for testing questions about group differences in ExWB.

Appropriate sampling, though, is necessary for estimates of group differences to be unbiased. One example of how inappropriate sampling could bias estimates occurs in the case where ExWB varies by day of the week and the groups to be compared are not sampled equally over days of the week. In this case, group differences would be confounded with day-of-the-week effects. Random sampling of SDMs over day-of-the-week is probably the best method for reducing the possibility of confounding, but stratified sampling strategies could be effective in smaller samples. (Appropriate data weighting methods may also be used to correct sampling bias.)

The Gallup datasets—along with others such as the International Social Survey Program, the World Values Survey, and the Survey of Health, Ageing and Retirement in Europe—have also afforded an opportunity to examine the statistical power (which is a function of daily variability) for the detection of between-group effects. This issue was addressed in Stone (2011) in which a survey of more than 300,000 individuals conducted by the Gallup Organization was analyzed. A small effect size of 0.11, statistical power of 0.80, and a two-tailed alpha level of 0.05 were used for determining the sample size necessary for detecting this magnitude of effect. A sample of only 2,796 people was necessary in this case (and these analyses were supplemented by simulations; see Stone, 2011), indicating that the very large sample for the original survey had extremely high power for detecting small effects with this "highly variable" daily measure (in this case it was a rating of the amount of stress experienced yesterday).

Another paper by Krueger and Schkade (2008) examined the test-retest reliability of SDMs derived from DRMs administered 2 weeks apart. They found that most ExWB indices yielded reliability coefficients between 0.50 and 0.70, with the multi-item scales for positive affect and negative affect having reliability coefficients of 0.68 and 0.60, respectively. Perhaps surprisingly, these reliabilities were in the same range as those for measures of life satisfaction (evaluative well-being). The authors concluded that, among other things, experience measures derived from the DRM are "sufficiently high to yield informative estimates for much of the research that is currently being undertaken on subjective well-being, particularly in cases where group means are being compared (e.g., rich vs. poor, employed vs. unemployed)" (Krueger and Schkade, 2008, p. 1843).

CONCLUSION 3.4: Although there may be an initial hesitancy by some to accept the utility of SDMs of ExWB because of their daily variability, a strong case can be made for their deployment in survey studies. Ideally, for a given respondent, capacity would be built into the survey design to aggregate over a number of days; controls for day-of-week effects need to be included in the survey design thoughtfully. In practice, multiday sampling will frequently not be possible; further,

government surveys will not always be the best option for carrying out this sort of detailed data collection. Sometimes, government-funded surveys and nongovernment data collections will possess a comparative advantage.

Another concern for interpretation of SDM data has to do with effect sizes. The unstandardized effect size (say a difference of 2 points between men and women on a 7-point scale) should be estimated in an unbiased manner regardless of the amount of noise in the SDM—that is, a 2-point difference by sex (for example) would be evident regardless of the variability of the measure, although in any single study the 2-point difference will be better approximated by a study with relatively lower variability. The same cannot be said for the standardized effect size that is often used as a measure of the strength of an association. Here, the noisy (higher variance) measure will have a lower standardized effect size. This distinction is important when comparing effect sizes from different measurement strategies, especially those that do not contain daily variation. Standardized effect sizes from SDMs will often be relatively small because of the daily variation.

There is also value to increasing the number of SDMs completed by each respondent. In the example above, taking the mean of 20 SDMs for each person would continue to yield a 2-point difference by gender (unstandardized effect size). However, the standardized effect size would be considerably higher because the variability inherent in one SDM will have been "averaged out" and the gender effect will appear relatively larger. Fewer participants would be required to achieve a given level of statistical power in this case. But the feasibility of administering multiple SDMs depends on a host of issues, including the relative burden for participants, the feasibility of implementation, and the costs to the investigator. Furthermore, the value in obtaining additional SDMs per respondent depends upon the amount of daily variability in SDM content for the population in question. Generally speaking, more is to be gained by adding additional assessments from an individual when there is much day-to-day fluctuation in the SDM.

Additionally, "simple" measures of ExWB, such as end-of-day and global-yesterday measures, which seem preferred for a broad range of surveys on practical grounds, need systematic experimentation that is informed by the extensive literature on retrospective versus concurrent reports of subjective experiences.

RECOMMENDATION 3.1 (Research): Despite the promising information available about SDMs, more information is needed about the psychometric properties of this class of ExWB measures. In particular, research is needed on how many days of data are generally needed to construct a reliable predictor (or average for a person) for end-of-day (or reconstructed yesterday) measures.

With respect to research into how many days might be needed to reduce within-person variability to tolerable levels or to identify the latent variable of ExWB, it should be noted that there cannot be one answer to the question that will be right for all people in all circumstances. Like correlation coefficients, the answer will be sensitive to a host of factors, so the most one study can do is provide a very rough estimate. Additional research is needed to further address such questions as "When is day-to-day variability itself of interest?" and "What do daily peaks and troughs in these data reveal?" For instance, some jobs create stress, which could be important. And various survey issues need more attention; for example, "What biases are created because working people are easier to reach on the weekends?" Additionally, because small effect sizes due to daily variation are compounded when the data are studied at the individual level (and most large-scale surveys are reported at the aggregate level where item reliability and effect sizes will be more substantial), measurement error differences between individual- and group-level measures should be investigated further.

Although some of this research could be carried out by statistical agencies prior to fielding a survey, most of this work will continue to be done by academic researchers working in the field, perhaps under research-grant programs supported by funding agencies.

RECOMMENDATION 3.2: For SDMs, day of sampling, time of day, and even respondent location, especially for certain subgroups, are important considerations when designing a study. These variables must be controlled for (which might just mean randomized sampling, so that the effects wash out) or avoided in measures of ExWB.

The above considerations are especially important for a statistical agency charged with developing and experimenting with single-day SWB questions.

3.3 RECONSTRUCTED ACTIVITY-BASED MEASURES

For some research and policy questions, contextual information about activities engaged in, specific behaviors, and proximate determinants is essential. For example, to investigate how people feel during job search activities, while undergoing medical procedures, or when engaged in child care, something more detailed than a global daily assessment is needed. Activity-based measures attempt to fill this measurement need. The attractive feature of activity-based measures is their capacity to improve understanding of the drivers of experience by providing dimensions of quality and context.

One promising activity-based ExWB measure is the DRM, developed by Kahneman et al. (2004). The DRM was created to assess subjective experiences in a manner that specifically avoids problems of many recall-based measures while being more affordable and less burdensome than momentary methods. The attractive feature of the DRM is its capacity to combine time-use information with the measurement of affective experiences. Respondents are asked to construct a diary of all activities they engaged in the preceding day; then they are given a list of positive and negative feelings and are asked to evaluate how strongly they felt each emotion during each activity listed in their diary, using a numeric rating (e.g., on a scale from 0 to 10). Participants follow a structured format in which they first divide a day into specific "episodes" or events. They then describe those events in terms of the type of activity (e.g., commuting to work, having a meal, exercising) and provide a detailed rating of their affective state during the activity.

Another attractive feature of DRM (and EMA) measures of ExWB is their potential to go beyond single indices of well-being that simply average across all ratings for an individual. While the mean certainly carries valuable information, it also ignores many other characteristics of experience, such as the amount of time spent in a particular hedonic state or the variability of hedonics throughout the day. Using DRM data, Kahneman et al. (2004) proposed the U-index, which is based on the relative intensity of positive and negative emotions during every episode; it yields a metric indicating the proportion of time respondents spent in predominantly positive or negative states. Thus, the richer data yielded by EMA and DRM have the potential to provide correspondingly deeper views of experience.

By asking participants to first recall the events of their day and then provide ratings associated with them, the DRM exploits the fact that, while memories of ongoing experiences such as pain and mood are flawed, memory for discrete events is more accurate (Robinson and Clore, 2002). Thus it avoids, or at least reduces, some of the biasing factors noted above, such as the tendency to recall information that is congruent with peak or recent experiences, which are more easily remembered. The DRM is designed to be self-administered and can be completed by most participants in a single sitting. It is thus much less burdensome and costly to field than the most rigorous EMA methods, and it is scalable to large surveys.

In assessing the value of DRM for estimating emotional experience, an obvious question is how well DRM results mirror those of more intensive methods such as EMA. Numerous concerns have arisen regarding the accuracy of traditional self-report measures that require respondents to remember and summarize their emotional experiences over some period of time or that ask respondents for on-the-spot judgments of their overall quality of life (for reviews, see Diener et al., 1999; Schwarz and Strack, 1999). These concerns have led methodologists to consider ways of capturing subjective experiences that rely less on participants' ability to remember subjective states accurately and to aggregate these experiences into a single summary score.

Among the issues remaining to be addressed about the scoring of DRM data is that many activity episodes are available for every person and many descriptive adjectives are available for each of the episodes. Apart from the U-index mentioned above, another attractive scoring method is to create a duration-weighted average of a selected adjective or composite of adjectives, where longer episodes contribute relatively more to the daily average than shorter episodes. This method has been employed in several DRM-type studies. However, for other purposes, different weighting schemes may be more appropriate, such as when high levels of the feelings of interest in a problem at hand are present; in this case, assigning higher weighting to episodes with feeling surpassing some threshold may be productive. Yet another option is to create metrics based on the content of activities; this may be appropriate if one were interested in well-being at the workplace or during specific activities. More research is needed to document the most efficient and useful ways to combine the rich information produced by the DRM.

3.3.1 Comparing DRM with Momentary Approaches

For ranking the relative merits of the competing ExWB measurement approaches, the panel took as its starting point the following statement, which is distilled from assessments of the reliability of SWB measures articulated by Krueger and Schkade (2008) and by Krueger et al. (2009). There is a compelling conceptual case for measures of ExWB (and, more narrowly, hedonic well-being) that is best satisfied by ESM/EMA, reasonably satisfied by the full DRM, and—with some compromises—sometimes satisfied by truncated versions of the DRM such as the ATUS SWB module.

This statement holds for cases in which momentary ExWB is the measurement objective. For some questions (e.g., predicting consumer behavior or whether or not a person is likely to repeat a medical procedure), a reconstructed assessment of ExWB may be more relevant;[5] it may also

---

[5] Posing the issue in a medical context clarifies the distinction. For instance, is the goal of a drug treatment to alter how much pain a person is in at a given moment or to alter how one remembers being in pain? The U.S. Food and Drug Administration tends to focus on actual pain; drug manufacturers may have a different objective.

be better at predicting a policy's impact on people's choices, but worse at assessing a policy's impact on experience. The director of a survey charged with adding self-reported well-being content has to answer the question, "Which ExWB approach should be used?" In many cases, the structure of the survey will rule out such approaches as EMA or end-of-day measures. For other predictive purposes, a cheaper evaluative well-being measure may perform as well as an ExWB measure.

This section discusses how momentary measures and reconstructed measures fare in terms of their relative susceptibility to context influences and the implications for accurate measurement. Overall, the panel concludes that episodic reconstruction can be quite accurate, at least for recent episodes, if respondents are given sufficient encouragement and time to relive the episode.

- *ESM/EMA* allow for introspective access to concurrent affect. By contrast, end-of-day and global-yesterday measures of ExWB require reconstruction; they differ in the extent to which they encourage and enable episodic reconstruction.
- *DRM—detailed reconstruction of yesterday.* A fully executed DRM encourages reconstruction of specific episodes, which is likely to induce a mild version of the affect associated with the episode. It captures EMA-like patterns that are not part of respondents' lay theories (Kahneman et al., 2004; Stone et al., 2006); this is important because it implies that the answers could not be produced by theory-driven reconstruction. A fully executed DRM takes considerable time, up to an hour in some cases, but Internet-based DRM versions may be more efficient. The time requirement precludes its routine use in representative surveys; even the reconstruction of partial days exceeds realistic resources for most studies.
- *Episodic with limited reconstruction of yesterday.* DRM adaptations with more limited reconstruction are more feasible and can reproduce core patterns obtained with the DRM and ESM. One version was implemented as the Princeton Affect and Time Survey; a variant is included in the ATUS SWB module, which assesses affect for three randomly selected episodes after respondents complete a whole-day stylized diary with minimal reconstruction of the three selected nonconsecutive episodes. (Notably, though, the entire day is reconstructed in both methods; only the feelings information is limited to being recalled for the selected episodes.)

The comparative properties of EMA and DRM measures of ExWB are a central concern for the future development of SWB survey modules. There are theoretical reasons, and some limited empirical evidence, to suggest that the DRM may provide some of the same advantages of EMA over traditional recall-based survey approaches. Several studies have directly tested whether the DRM can be used instead of EMA in some research contexts. Box 3-1 relates findings of a recent study that directly compares DRM and EMA data for the same participants during a given time period. The approach was to collect EMA, DRM, and standard recall-based measures of mood and physical symptoms in older adults, many of whom have osteoarthritis.[6] EMA-based ratings of emotional experiences as they occurred throughout the day, a reconstruction of those experiences using the DRM, and memory-based estimates from standard survey items were all obtained. If the DRM measure provides a close replication of actual experiences, one would expect to see high concordance between the DRM and EMA measures. If, on the other hand, DRM estimates are biased due to their reliance on recall, they should more closely match the estimates based on standard recall-based measures.

The findings in Box 3-1 suggest that DRM measures of mood and physical symptoms closely approximate summary measures created from an EMA protocol. Where there were systematic differences, DRM estimates of negative mood and physical symptoms, such as pain and fatigue, tended to be lower than those collected by EMA. In terms of within-day patterns, the correspondence between EMA and DRM estimates was striking; furthermore, both estimates diverged from participants' expressed beliefs about those patterns. The investigators also noted what appear to be advantages of the DRM measures over traditional recall-based summary measures, even with a time frame for the recall measures (4 days) that is shorter than usual.

> CONCLUSION 3.5: Preliminary assessment of DRM measures of mood and physical symptoms suggests that they reasonably approximate summary measures created from EMA protocols. An attractive feature for survey objectives is that the DRM approach goes beyond simply addressing who in the surveyed population is happy to identifying when they are happy. Additionally, it appears that the DRM is less burdensome on respondents than experience sampling, and it might reduce memory biases that are inherent in global recall of feelings. The DRM is thus a promising method for assessing feelings, mood, and physical symptoms that accompany situations and activities more

---

[6] Because this study was based on a sample of people with osteoarthritis, the associations observed may be somewhat higher than in other samples because of the relatively high variability of pain and fatigue likely in this group and therefore may not generalize to the population at large. On the other hand, one potential problem in comparisons between EMA and DRM data of this kind is that, if there is little within-person variation, the strength of associations between different measures will be low even if the measures themselves are relatively accurate.

BOX 3-1
A Test Comparison of EMA and DRM Estimates

Smith et al. (2012)* surveyed 120 older adults (age > 50), 80 of whom have osteoarthritis of the knee. These participants completed an EMA protocol over 4 days. It used a fixed interval schedule with 6 prompts per day (upon waking, 2, 4, 8, and 12 hours after waking, bedtime). Patients were asked about their mood (happiness, depression), symptoms (pain and fatigue), and level of physical activity. Because of the focus on transient physical symptoms in a smaller clinical sample, the investigators used a "coverage model" for the EMA protocol. That is, they asked participants to summarize their mood and symptom levels *since the last prompt*. On one day, participants also completed an Internet-based version of the DRM, which included the same measures (with identical wordings and scales) as the EMA protocol. The DRM protocol asked about activities and feelings for the previous day; and therefore was administered on day 2, 3, 4, or 5 to correspond to EMA day 1, 2, 3, or 4, respectively. On day 5, participants completed the same measures in summary form (e.g., "Over the past 4 days, how happy were you?").

With the data from this design, the investigators created overlapping EMA, DRM, and standard recall measures. A key difference is that the DRM sampled only one day. To allow comparison with the standard summary measures, the investigators created composite estimates of each measure by averaging responses to all 24 EMA prompts and all DRM activity ratings. Thus, each participant has one EMA score for average happiness, one DRM score for average happiness, and of course the single summary score from the standard recall-based measure.

Mean Levels of Mood and Symptoms by Method of Assessment

	EMA Estimate	DRM Estimate	Recall-Based Estimate
Happy	2.8	2.7	3.0
Depressed	0.6	0.5	0.7
Pain	1.2	1.1	1.5
Fatigue	1.3	1.1	1.4
Activity level	1.8	1.6	2.7

Compared to the EMA estimates, standard recall-based survey measures of happiness, depression, pain, fatigue, and activity level all showed levels that were higher, and markedly so in the case of activity level (all comparisons significant at $p < .05$, with the exception of fatigue; Stone et al., 2012, p. 10). This pattern is consistent with memory estimates that were biased by "peak" experiences (Broderick et al., 2009). Person-level correlations between the recall-based and EMA estimates were strong, ranging from $r = 0.53$ (activity level) to $r = 0.86$ (physical pain); average correlation across all five measures was $r = 0.75$. In contrast, DRM levels of happiness were nearly identical to those of EMA ($p = 0.24$). Estimates of pain, depression, fatigue, and activity level were all slightly lower, which is not consistent with a peak bias (all p's < 0.05). In addition, person-level correlations were higher

> **BOX 3-1 Continued**
>
> than those observed with the recall-based measure in every instance, ranging from $r = 0.65$ (activity level) to $r = 0.92$ (physical pain); average correlation across all five measures was $r = 0.81$.
>
> Interpretation of these comparisons is somewhat complicated by the fact that the EMA and recall-based measures cover 4 days, compared to 1 day for the DRM measures. Thus, the investigators restricted the EMA time range to the single day on which the DRM was completed, but this made little difference; the means were again similar across the EMA and DRM methods, and the correlations were nearly identical.
>
> Diurnal patterns were examined next. Both EMA and DRM revealed similar cross-day changes in mood and symptom levels. When participants were asked to estimate how their mood and symptom levels typically changed throughout the day, they did not reproduce these patterns (with the notable exception of physical pain). Thus, they appeared to be mostly unaware of the patterns present in the scores they had provided over the previous 4 days. This analysis is important, because it shows that DRM diurnal patterns resembled EMA patterns more closely than they resembled participants' *beliefs* about these patterns. Where the recall-based measure diverged from the EMA averaged estimate, the DRM measures still tracked with the EMA measures.
>
> ---
>
> *This summary of findings was commissioned by the panel and funded by the National Institute on Aging. Susan Murphy, Norbert Schwarz, and Peter Ubel, as well as study team members William Lopez and Rachel Tocco, were study co-investigators with Dylan Smith.

efficiently than with EMA methods and with greater specificity and accuracy than traditional recall-based methods.

While the DRM is certainly promising, questions have been raised about its use—for example, about the extent to which estimates produced are unbiased and about whether the time weighting implied in the survey structure reflects psychological realities (see Diener and Tay, 2013). Thus, the panel adds to the above conclusion the caveat that research using the DRM is still in an early stage, so evidence for the validity and reliability of the DRM is, at this time, somewhat limited. This constraint suggests the need for further comparative research using ESM/EMA data to validate the DRM.

RECOMMENDATION 3.3 (Research): Additional research is needed to better establish the evidence base for determining when the DRM is an adequate substitute for EMA methods of measuring ExWB. In

particular, better understanding is needed of the psychometric properties of the DRM; this may be achieved, for example, by comparing DRM reports to mobile phone assessments and other forms of momentary experience sampling, as well as to global reports of feelings in situations. Additionally, more research is needed comparing performance, sensitivity, and variation of DRM and EMA approaches to measuring ExWB.

For some purposes, the DRM will not be an adequate substitute for momentary experience sampling. While it is possible to learn a great deal about people's emotional states associated with various activities using the DRM, additional work is needed on the meaning or interpretation of self-reports summarized over a period of time.

3.3.2 Time-Use Surveys

The American Time Use Survey (ATUS), conducted by the Bureau of Labor Statistics, included an SWB module in 2010 and 2012, which was funded by the National Institute on Aging. The ATUS SWB module, which is described in detail in Appendix B, is the only federal government data source of its kind, linking self-reported information on individuals' well-being to their activities and time use. The ATUS SWB module is thus an abbreviated version of a DRM approach. There are other short-form versions of the DRM that have been used in experimentation, such as the Princeton Affect and Time Survey, mentioned above.

Much of the policy promise of ExWB measurement lies in its potential to be combined with time-use information designed to illuminate how activities and environments relate to a person's emotional states. If activity and time allocation are not included in a survey design, data analyses are limited to considering the influence of sociodemographic characteristics, such as those that dominate the literature on evaluative well-being.

CONCLUSION 3.6: Capturing the time-use and activity details of survey respondents enhances the policy relevance of ExWB measures by embedding information about relationships between emotional states and specific activities of daily life.

It is a relatively easy task to identify examples where detailed time-use survey data add analytic content beyond that which is obtainable from global-yesterday measures:

- Commuting effects cannot be teased out of global-day measures. In contrast, Christmas encompasses a day, so a yesterday measure will in principle call attention to effects associated with it.
- Tracking the health of a population may not require detailed activity-based data. But to get at causes of stress or even pain, researchers need data on the activities associated with these affects.
- Using an overall day measure, an unemployed person may look only a little worse off (or not at all) than the population average. Analysis needs to look at differentials at work and during activities while not at work; otherwise any explanation of the self-reported results is incomplete.
- In terms of policy pathways, time-use data provide insights into what income is a proxy for. Such data capture effects on emotional states of being on vacation, enjoying leisure, being at work, etc.

How well the ATUS truncated version of the DRM will ultimately perform is yet to be determined. However, it is not too early to begin taking advantage of the opportunity afforded by the ATUS SWB module to explore this issue.

> **RECOMMENDATION 3.4:** While it may not be practical to run the ATUS as a full DRM—although this would yield very valuable information—it may be possible to explore differences between the ATUS SWB module and a full DRM by using a pilot consisting of a sample of ATUS respondents. In addition, increasing the number of episodes examined for ExWB would be desirable.

More generally, for DRM-type survey designs, much more can be learned when survey modules are placed so that samples are drawn from people for whom much is already known—e.g., subsamples of the Understanding Society Survey, Current Population Survey, and others that are rich in relevant covariates. Among additional research questions, one is how to weight events in a DRM approach, given that people experience different numbers of episodes of different durations and that affect has been shown to correlate with duration of episode. Another key research question is the reliability and usefulness of shorter, hybrid, DRM-like methods linking to activities.[7] The overall goal of this research would be to produce something

---

[7] In research being funded by the National Institute on Aging, Jacqui Smith and colleagues are tackling this issue by comparing Health and Retirement Study findings with the DRM data collected in the Panel Study of Income Dynamics, the ATUS SWB module, and the American Life Panel's DRM measures. These secondary analyses will answer questions about the quality and comparability of responses to the fine-grained DRM approach versus brief DRM measures. Available: http://micda.psc.isr.umich.edu/project/detail/35382 [October 2013].

better (more information content) than simple overall day measures, while still being short enough in administration time required to add to surveys with minimal increase in respondent burden.

RECOMMENDATION 3.5 (Research): Additional research is needed on the optimal response scales and on the various ways of creating summary measures of the day's affect. Although duration-weighted measures are usually used, other combinations of the data from time-use and affective data are possible, such as the U-index.

Chapter 6, on data collection strategies, returns to considerations about the next steps for the ATUS SWB module and other shortened variants of the DRM.

4

Additional Conceptual and Measurement Issues

This chapter continues the discussion of experienced well-being (ExWB) measurement, addressing some specific issues that have arisen as the research base has evolved and grown. The following issues are discussed here:

- Whether respondents' answers to ExWB questions are subject to systematic biases and differences between groups—defined by culture, age, or other traits—that may invite misleading conclusions about respondents' actual hedonic experiences;
- Susceptibility of ExWB measures to various biases induced by context or by question ordering, and the importance of these effects;
- Sensitivity of self-reported ExWB to changing situations and environments;
- The role of adaptation and response shift in ExWB measurement; and
- Scale and survey mode effects and the design of instruments.

4.1 CULTURAL CONSIDERATIONS

The value that people place on various emotional states shapes their reports of subjective well-being (SWB). A large body of research shows systematic variations in self-reported well-being that appear to be associated with cultural norms about ideal affective states (see Tsai et al., 2006, for a review). Consequently, when making international comparisons or interpreting findings from various subpopulations within a country, care must be taken to consider cultural contexts. Asians and Asian Americans, for example, appear to place less value on excitement and joy than on states characterized by calmness and serenity. In contrast, European Americans are exceptional in the considerable value they place on high-arousal positive states such as excitement and surprise. Such observations can raise questions, even doubts, about the meaning of comparisons of SWB across countries. The issue is obviously important in measurement, and because SWB is a topic frequently discussed in the media, it is also important when communicating findings to the public.

Arousal appears to be a key dimension that distinguishes subgroups. East Asians, as well as older people, tend to endorse more low-arousal positive emotions than high-arousal positive emotions (Kessler and Staudinger, 2009; Tsai, 2007; Tsai et al., 2006). As noted above, East Asians place less value on surgency than do westerners (Tsai, 2007; Tsai et al., 2006).[1]

Tsai et al. (2006) argue convincingly that despite great cultural consistency in the subjective and physiological experience of emotions once they are *elicited*, cultures vary considerably in how people *want* to feel. Anger and sadness appear to be more acceptable states among Germans than Americans, for example. Similarly, at older ages people report more mixed emotional experiences, even though they report higher overall levels of SWB than younger adults (Ersner-Hershfield et al., 2008). Moreover, mixed emotional experience is associated prospectively with better physical health across adulthood (Hershfield et al., 2013), suggesting that mixed emotions do not detract from SWB in older populations. More research is needed on ethnic and age differences in affect valuation, especially in the United States, where ethnic diversity is increasing. Under mainstream assumptions about immigration policy and trends, Hispanics will account for two-thirds of the growth in U.S. population from 2010 to 2050, and the proportion of older people will increase from 13 percent currently to 20 percent in 2030 (Passel and Cohn, 2008).

Despite such variations in the factors that contribute to it, *happiness* itself appears to be understood in much the same way across cultures. Thus, SWB measures based on "happy yesterday" questions may be especially useful for international comparisons because they offset the fact that different factors (e.g., arousal or calm) may contribute to subjective happiness at different ages or in different cultures.[2]

---

[1] Indeed, the basis for self-reported measures may even change for a given individual. Oishi et al. (2003) found that excitement was more often a factor in life-satisfaction judgment on weekends than on weekdays.

[2] Fulmer et al. (2010) showed that people are happier when their personalities match their cultures. That is, the extent to which people's personalities, for example, traits of the "big five" personality theory, predict their SWB and self-esteem depends on the degree of personality match to the dominant personality dimensions in the culture.

RECOMMENDATION 4.1 (Research): More study is needed about the role of cultural effects on ExWB. In particular, the value placed on high-arousal positive states versus low-arousal positive states and the acceptance of negative states, like anger and sadness, likely varies considerably by age and cultural context, which suggests that subpopulations assess ExWB differently. For example, if a measure relies heavily on high-arousal positive items, older populations will appear less happy; a similar bias may occur in assessing some Asian populations.

The use of anchoring vignettes is a promising approach to identifying and correcting systematic cross-cultural differences in question interpretation. Such approaches have been used in a number of contexts, such as in cross-country comparisons of job satisfaction (Kristensen and Johansson, 2008) or life satisfaction (Kapteyn et al., 2010). Van Soeste et al. (2011, p. 575), in an assessment of this growing literature, conclude that "vignette based corrections appear quite effective in bringing objective and subjective measures closer together." Notwithstanding this promising beginning, the approach's effectiveness will not be fully assessable until further research is conducted in a range of contexts and on a range of outcomes.

4.2 AGING AND THE POSITIVITY EFFECT

Because of its key research policy interest, attention to aging as it affects memory for emotional experience merits consideration in the measurement of ExWB. The *positivity effect* refers to an age-related trend that favors positive over negative stimuli in cognitive processing. Relative to their younger counterparts, older people attend to and remember more positive than negative information.

The positivity effect has been documented across a variety of experimental paradigms and a wide range of stimuli, supporting the robustness of the effect (Reed and Carstensen, 2012). It emerges in studies of working memory (Mikels et al., 2005), short-term memory (Charles et al., 2003), autobiographical memory (Kennedy et al., 2004; Schlagman et al., 2006), and even false memories (Fernandes et al., 2008). It is also evident in decision making. Compared to younger people, older people pay greater attention to positive as compared to negative attributes when, for example, choosing doctors and hospitals (Löckenhoff and Carstensen, 2007, 2008) and making decisions about consumer products (Kim et al., 2008). Compared to younger adults, older adults also remember their choices in a manner that is positively skewed—either via disproportionately recalling positive attributes or via attributing positive attributes to chosen options and negative attributes to rejected options (Löckenhoff and Carstensen, 2007, 2008; Mather and Johnson, 2000; Mather et al., 2005).

The effect appears to reflect a top-down, motivated process in which cognition operates in the service of affect regulation. That is, there are changes in cognitive processing associated with age-related changes in goals that prioritize emotional satisfaction and meaning (Carstensen, 2006). Positivity is most evident in automatic (impulsive) processing and less so in deliberative processing (which entails cognitive work); indeed, experiments that emphasize attention to detail eliminate the effect (Löckenhoff and Carstensen, 2007). Thus, although empirical examination is needed, the deliberative processing inherent in the DRM would likely reduce or eliminate age differences in that it involves reflection.

4.3 SENSITIVITY OF ExWB MEASURES TO CHANGING CONDITIONS

Among the most crucial issues for ExWB measures are their ability to distinguish groups or sectors of the population and their sensitivity to change. An additional and as yet unanswered question relevant to assessments of their applicability to policy is what constitutes a meaningful change in ExWB measures. Assessing significance ("meaningfulness") is an obvious challenge given that these are subjective variables that run on an ordinal scale and that there may be (statistically) significant differences in terms of what is meaningful over time versus across cohorts in a cross-section (and there is likely more margin of error in determining the latter). While there is no single answer to this question, ExWB measures should not be held to an unachievable standard—for example, one that is higher than the standards set for other dimensions of social and economic measurement.

To influence long-term ExWB substantially at the population level, government policies designed to change the everyday circumstances of individuals would have to affect very large groups of people (as they sometimes do) on a day-to-day basis. Socially traumatic events like the assassination of President Kennedy or the 2001 terrorist attacks, have had a detectable impact on measures of ExWB at the national level, but the measured effects have typically been short-lived. This highlights an important difference between evaluative well-being and ExWB: the latter primarily reflects what is currently engaging peoples' attention; much less so events from the past, even important ones. (After a major event, the immediate environment, like being engaged in family or work activities, may be able "to grab" a person's attention and influence their ExWB.) Even a large increase in unemployment, such as accompanied the severe recession of 2007-2011, may have only a muted impact on response means of SWB measures when the change in unemployment takes place over a number of months and directly affects only a small percentage of the population.[3]

In addition to policy issues couched at the macro level, sensitivity to change is relevant to measures most likely to be useful at more local levels—for example, to assess the impact of local initiatives such as changes to traffic management, crime programs, or local school policies. This relates to the issue of targeted versus general measures. For example, if one wants to know about the impact of a traffic management policy or a health care innovation, then measures that specifically target people's experience of traffic or health will likely be more sensitive than general well-being assessments.

In thinking about how to calibrate ExWB measures to address sensitivity concerns, it is instructive to think of examples of change in other statistical constructs, such as unemployment or income change. The unemployment rate rarely changes quickly, and a change from 6 to 6.1 percent reflects a change in status of only 1 in 1,000 people in the work force. Over the 50 years of existence of national unemployment statistics, economists have had time to learn how to understand and interpret what appears to be a small change; for example, the change from 6 to 6.1 percent represents a much larger impact among the population defined as actively looking for work. At present, the time series of SWB data is of insufficient length to instill confidence that it does or does not move over time or to know how to interpret a change as a small versus big movement.

Changes in income are often benchmarked against changes relative to a peer or professional group or against some threshold such as the poverty line. If one assumes a curvilinear relationship between income and happiness (that is, the widely supported generalization of decreasing marginal utility for higher levels of income), then a positive change in income will have less effect on ExWB than a negative change of the same percentage. But the exact relationship is debated, and, as Easterlin (2005, pp. 252-253) points out, "the cross-sectional relationship is not necessarily a trustworthy guide to experience over time or to inferences about policy." Given these uncertainties, the answer as to what constitutes a meaningful change in income could be informed as much by SWB metrics as by income metrics. It is possible to measure the effects of these changes—and their relationship to changes or lack thereof in relevant cohorts—on SWB in a way that cannot be captured by revealed preferences.[4]

---

[3] To be clear, life evaluation (evaluative well-being) measures may trend quite differently. The generalization made by Stiglitz et al. (2009) or the *World Happiness Report* (Helliwell et al., 2012) about the high human costs of unemployment is based on life-evaluation measures, not measures of momentary emotional states.

[4] Economists have generally relied on revealed preferences—observations of people's actual decisions and choices—as opposed to self-reports of intentions or inclinations. The opening

In the case of some ExWB measures, obvious thresholds exist—a move from a positive to a negative self-assessment, for example. And reducing suffering is surely a meaningful goal. In the same way that targeting the needs of the poor is an important but not the only objective of macroeconomic policy, prioritizing the needs of those in misery is one possible objective of policy informed by ExWB metrics. To the extent policies aim to increase the capabilities and opportunities of the most number of citizens possible, then some attention to increasing SWB as measured by measures of eudaimonic well-being may also enter into policy priorities. Ultimately, a discussion of what the priorities are would help establish what constitutes a meaningful change, or at least establish parameters for assessing changes in the aggregate and those that affect particular cohorts. We have already concluded that aggregate tracking is not what ExWB measures are likely to be most useful for. But can changes in ExWB be reliably detected at the individual level, or is it more realistic and useful to attempt measures for population groups? Research attention is needed to strengthen the evidence base for addressing these and related questions.

Additionally, the temporal nature of change has implications for the kinds of datasets needed. A different data collection approach—e.g., how often people are surveyed—is implied for measures that are sensitive typically on a very short time frame, responding to daily events, versus those that move very slowly. If data are collected every 2 years on a large survey, they are unlikely to be capable of catching short-lived deviations in ExWB (such as those associated with weekends or holidays). Such trends may need to be assessed using higher-frequency data collections with smaller samples, as opposed to massive population surveys conducted annually or every several years. Consumer confidence may be an example of this kind of rapid-change pattern, which may explain why the survey on which the University of Michigan Consumer Sentiment Index is based uses fairly small samples but an ongoing data collection design. On the other hand, large samples may be needed to inform macroeconomic policies about the broad

---

paragraphs of Kahneman and Krueger (2006) identify some of the respective roles for and strengths and weaknesses of SWB and revealed preference approaches. Fujiwara and Campbell (2011) provided a detailed assessment of valuation techniques—specifically, those based on revealed preference, stated preference, and SWB methods—for estimating costs and benefits of social policies. One of their conclusions was that while, at the moment, SWB methods often yield implausible estimates (as do revealed preferences in many cases), "they may still be useful in challenging decision makers to think more carefully about the full range of impacts of their proposed policies. And they may help decision makers to question the values that they may otherwise place implicitly on these impacts" (Fujiwara and Campbell, 2011, p. 53). Dolan and Metcalfe (2008) compared individuals' willingness to pay for goods and services related to urban regeneration using revealed preference and SWB methods. They found "that monetary estimates from SWB data are significantly higher than from revealed and stated preference data" and explain possible sources of these differences.

population impact of factors such as unemployment or inflation, which themselves do not often change quickly, or to identify outlier populations that are suffering substantially more than the general population.

If ExWB measures are to be used meaningfully, data users need to know something about how to interpret changes in their value. For example, on a scale from 0 to 10, how does a change from 1 to 2 compare to a change from 7 to 8? At a minimum, it is preferable for the range of values to lie on an interval scale such that each increment on the scale is valued equally. As an alternative, an adjustment factor that accounts for nonlinearities (e.g., end-point aversion) could be applied to the change in rating, but it is unclear what the adjustments should look like. If one can only say that 2 is better than 1 but not by how much, it would only be possible to use ExWB measures as ordinal representations of value; this would seriously limit their applicability. That said, ordinal data could be combined with duration to calculate the percentage of "unhappy" time over the day. This is the approach adopted by Kahneman and Krueger (2006) in calculating the U-index using data from the DRM. But this approach loses potentially important information about just how bad the "unhappy" time is (and just how good the remaining "happy" time is). And it assumes that feelings of relative goodness and badness are independent of, and linearly weighted by, their duration. People care about being happier for longer, but the SWB research field has not made much progress on methods for comparing "how much happier" with "how much longer."

4.4 ADAPTATION, RESPONSE SHIFT, AND THE VALIDITY OF ExWB MEASURES

Hedonic adaptation is the psychological process whereby people adjust to and become accustomed to a positive or negative stimulus brought on by changed circumstances, a single event, or a recurring event. People's responses to questions about their well-being or quality of life have often reflected this, which potentially poses problems for using SWB measures, particularly for sorting out longitudinal effects when multiple determinants are at work.

Interpreting "response shifts," a term used to characterize change in reporting over time, is complicated by the possibility that observed differences over time in self-reports of well-being may reflect true change in a respondent's quality-of-life assessment (e.g., hedonic adaptation); measurement error (e.g., that associated with "scale recalibration" bias); or both. Ubel et al. (2010, p. 466) provided the following hypothetical examples:

1. A person's happiness is partially restored after paraplegia. Over time, reported mood improves as the person begins shifting his

focus away from what he cannot do and toward new goals (e.g., from jogging to participating in wheelchair basketball). The individual's physical functioning does not improve or deteriorate but, over time, his or her responses to well-being questions shift as the percentage of time experiencing positive emotions increases and the percentage of time experiencing negative emotions decreases.
2. A person with chronic pain experiences kidney stones. Prior to the bout with kidney stones, the person rates her chronic pain a 7 out of 10, on average. Then, she experiences kidney stones for which the pain is much more intense. This episode leads her to reinterpret the pain scale, and she shifts her response to now rate the (unchanged) chronic pain at only 5 out of 10.

The reported scores of the case 1 person reflect a true change in ExWB that occurred as a result of hedonic adaptation or a change in values; case 2, in contrast, does not provide a valid assessment of the person's pain levels over time but is simply a recalibration of the reporting scale.[5] For most purposes, researchers are interested in isolating the first category of phenomena, without the potential confounding of the second type of response shift.

Another example of scale recalibration has to do with how questions are interpreted. People may normalize their responses to questions about experienced utility (or other dimensions of SWB) to implicit standards of comparison (Kahneman and Miller, 1986). For example, people who have experienced a decline in functioning may norm their responses relative to their perceived assessment of others with the same disability. People may also reconceptualize SWB questions. For example, after surviving a cancer scare, a person may reprioritize and become more concerned about engaging in meaningful activities as opposed to immediately enjoyable ones (or vice versa). This creates a measurement issue in applications for which a consistent definition of SWB (or one dimension of it, such as ExWB) over time is required; if the measurement objective allows for individual interpretation of the SWB construct of interest, then this reconceptualization may not be an issue. Ubel et al. (2010) argued that shifts in actual well-being (adaptation) and scale recalibration are distinct causes of response shift and need to be disentangled; they proposed doing away with "response shift" terminology because of this ambiguity.

Much of the relevant research on response shift—and components thereof—is in the health care/clinical trial literature and has been conducted to more accurately assess quality of life among the chronically ill or the

---

[5] It may be possible that having experienced the more severe pain allows the person to cope with the chronic pain with a new perspective and less distress. In other words, the measurement of pain may not be valid, but measurement of SWB may truly have shifted (as in case 1).

disabled, as in the following two examples. On the topic of cognitive adaptation, the 1978 paper by Brickman et al. on lottery winners and long-term paraplegics was highly influential in establishing the idea that, after these events, reported life satisfaction of the affected individuals returns to pre-event levels more quickly and completely than would be expected either intuitively or by people predicting what their moods would be under those conditions. For ExWB, one possible explanation of hedonic adaptation is offered by the set-point theory, which posits that people initially react to events, but then return to some baseline that is determined by personality factors (Brickman and Campbell, 1971).

Research subsequent to Brickman et al. (1978), much of which has used longitudinal data, has shown that adaptation is more complex than portrayed by some of the earlier studies and not as universal as once thought. The extent to which adaptation occurs may vary a great deal, depending on the exact nature of the event or circumstance that alters SWB. For example, the temporal impact of marriage on SWB, including affect, appears to often be short-lived (Clark et al., 2008), while the effects associated with unemployment and chronic pain appear to be more long-lasting (Lucas et al., 2004).

Loewenstein and Ubel (2008) measured the moment-to-moment mood of healthy people and dialysis patients over the course of a week; they found only small differences in the level of positive and negative mood recorded by the two groups. In other words, the dialysis patients presumably experienced a significant amount of emotional adaptation to their illness. Riis et al. (2005) documented patterns of adaptation and underprediction of adaptation when eliciting momentary measures of ExWB. Spikes of grief associated with loss of a child, say, may not show up in experience sampling methods. A separate question is how to account for the intensity of these kinds of emotions. Analyses by Diener and colleagues (1999) and others corroborate the conclusion that experience and evaluation dimensions of SWB trend differently after a life-changing event in terms of extent and pace of adaptation. Using data from the Household Income and Labour Dynamics of Australia survey, a long-term longitudinal panel study, he found that disability cases showed approximately the same pattern of decrease in positive effect and increase in negative effect, with very little adaptation. For the loss of spouse or child, negative feelings increase sharply after the death but then fully return to baseline, while positive feelings rebound some, but do not return to previous levels (Clark et al., 2008).

CONCLUSION 4.1: The evidence with regard to adaptation suggests that it cannot be characterized as a process that occurs uniformly; people adapt differently to different events and life changes, in some part due to norms and expectations. Ideally, question structures should

be designed to allow researchers to decompose changes in response scores into scale recalibration (or other measurement errors) and true quality-of-life change components.

For example, in hypothetical case 2 above from Ubel et al. (2010), the person could have been asked to rate both quality of life and pain, which might allow the separate effects to be teased out.

In terms of its effect on policy relevance, if reported SWB (either ExWB or evaluative well-being) were not closely linked with individuals' circumstances and opportunities, due to adaptation, the question arises of whether the measures are exploitable to inform policy. Smith et al. (2006) found that people report a willingness to pay large sums of money or make other major sacrifices to restore lost functions. Loewenstein and Ubel (2008, p. 1797) wrote:

> A key problem with experience utility as a welfare criterion for public policy is its failure to sufficiently value negative or positive outcomes that people adapt to emotionally. It is well documented that people exhibit near-normal levels of happiness not long after experiencing adverse outcomes such as paraplegia, colostomy or end-stage kidney disease. Yet, the same people often report a willingness to make great sacrifices to alleviate their condition. A welfare criterion based on experience utility would run the risk of failing to treat such outcomes as welfare-diminishing—e.g., of treating an increase in cases of paraplegia as a welfare-neutral event.

A broad implication of this line of thinking is that policy makers should be aware that people care about aspects of their life that cannot be captured by a single measure, whether it is willingness to pay, experienced utility, or something else. Multiple kinds of evidence need to be considered. Loewenstein and Ubel (2008) suggest that, given limitations of both decision utility (based on ordinal utility concepts) and of experienced utility measures, evaluations of welfare will inevitably have to be informed by a combination of both approaches, patched together in a fashion that will depend on the specific context. The goal of policies ought to be to maximize people's SWB, but the moment-to-moment or ExWB dimension is only one component of SWB.

Variation in the extent to which adaptation occurs in response to different domains, conditions, or cases may actually convey a great deal of information that is relevant to policy. For example, information about how people respond and adapt to price inflation versus unemployment, or the threat of it (Di Tella et al., 2001), would seem highly relevant to policy. The same may be true for data on how people who become severely disabled from a job-related accident respond or adapt differentially to psychological scarring and emotional harm caused by inability to continue working and to a compensation package for lost income. In the health care context, Dolan and Kahneman (2008, p. 221) concluded that:

> in general, it seems entirely appropriate [for ranking policy options] to give greater priority to those states that people do not adapt to over those that they do adapt to. This would seem to be particularly true when allocating resources amongst patients once the budget for health care has been determined i.e., once we have decided the priority afforded to patients in relation to other groups. Given this, we need to consider how well people predict changes—including any adaptation—in their future preferences.

Furthermore, people compensate for (adapt to) having poor education, to living in poverty or high crime areas, etc., yet these are certainly important policy areas. Understanding why people tolerate poor norms of health or lots of crime and corruption or bad environments seems especially relevant (Graham, 2011; Sen, 1985).[6]

4.5 SURVEY CONTEXTUAL INFLUENCES

All human judgment is subject to contextual influences, and the same holds for all self-reports that serve as measures of SWB. Focusing effects have a big impact. When asking across people, it is difficult to know if context is a biasing factor and, if so, for whom. How problematic a given contextual influence is depends on the objective of the measure. These objectives differ across measures of SWB, which renders various types of contextual influences differentially problematic.

Evaluative well-being involves assessments of extended periods of time, often a respondent's "life-as-a-whole" or "life-these-days." Such questions explicitly ask respondents to include all aspects of life (or the respective life-domain)—for example, "Taking all things together. . . ." If this is the goal, *any* transient influence on judgment represents undue contamination in the form of giving too much weight at the moment to things the respondent would consider irrelevant if asked about them specifically. Typical examples include naturalistic context variables (e.g., the weather at the time of interview, sports news of the day) and research instrument variables (e.g., question order). For a comprehensive review, see Schwarz and Strack (1999).

In contrast, measures of ExWB attempt to assess how respondents feel during a much shorter reference period or episode, on which respondents

---

[6]There are, as discussed in section 4.4, two distinct influences working here, which should not be confused: adaptation to conditions and cognitive states reflecting low expectations.

may report either immediately or retrospectively (see Chapter 3). In the ideal case, ExWB is assessed through concurrent reports of affect in situ (that is, with momentary assessment methods as discussed in section 3.1). Concurrent reports allow for introspective access to one's momentary feelings and are the ideal option for their assessment (Robinson and Clore, 2002). Under such conditions, temporary influences arising from the context of daily life do not represent undue contamination; those whose moods were lifted by sunny weather or news about a sports event did indeed experience a period of higher ExWB. That such events are reflected in ExWB measures is testimony to their sensitivity, whereas the same influence would constitute a source of context bias for measures of evaluative well-being; people usually assume that good news about a favorite sports team can brighten one's afternoon for a couple hours but not improve one's life-as-a-whole, "taking all things together." In contrast, when ExWB is not assessed immediately, temporary real-life influences at the time of measurement can bias retrospective reports. Accordingly, different measures of ExWB differ in their susceptibility to bias.

Finally, the influence of research instrument variables always presents undesirable contamination on measures of SWB, whether they pertain to ExWB or evaluative well-being. When the goal is to draw conclusions about a population, any influence that merely affects the sample and was not part of the experience of the population undermines the purpose of the assessment.

Psychological research shows that many feelings are fleeting. An individual can introspect on them while they are occurring (making Ecological Momentary Assessment the gold standard for ExWB assessment) but will need to reconstruct them after they have dissipated (Robinson and Clore, 2002; Schwarz et al., 2009). The extent to which the reconstruction captures the actual experience depends on the temporal distance between the experience and the time of interview and the extent to which respondents "relive" the past experience prior to reporting on it (i.e., the extent to which they reinstantiate the experience in memory). Thus, the potential for bias is likely to increase with the length of the episode and its temporal distance from the interview, and to decrease with the detailed reinstantiation of the episode. The available data are compatible with these assumptions, but more systematic comparisons across measures, based on the same population and time frame, are needed.

Considerations of context effects have played a strong role in the conceptualization and development of ExWB measures, and that work is continuing to explore the effects of context and determine ways to reduce unwanted effects. Several effects specifically related to context are discussed in the following sections.

4.6 QUESTION-ORDER EFFECTS

Assessments of SWB, both evaluative and experienced, typically depend on the administration of several questions. In the case of ExWB, these are often questions about a series of adjectives, posed to the respondent one after another. One concern is that the order of the questions or the order in which adjectives are presented may introduce random error or, worse, bias in the ExWB measures. In addition to the order of questions within an assessment, there is evidence (discussed next) that the content of questions that precede an ExWB assessment may influence the answers. The nature of these effects, and their directionality and magnitude are important considerations in the design of SWB research protocols.

There is a large literature on question context and order effects.[7] Schimmack and Oishi (2005), in a review of 16 studies, found that only 3 of them exhibited significant item-order effects. The authors concluded that order effects are often unimportant in actual survey settings because (as summarized by Diener et al., 2013, pp. 13-14), "chronically accessible information is not raised in importance by priming because it is already highly accessible, and other information is often ignored because it is seen as not relevant." Other investigations raise serious concerns however. A split-sample randomized trial using experimental national data conducted by the UK Office for National Statistics (ONS) reported an effect of question order on multiple-item positive and negative emotion questions (Office for National Statistics, 2011). Asking negative emotion questions first produced lower scores on some positive emotion items for the adjectives "relaxed," "calm," "excited," and "energized." When positive emotion questions were asked first, the mean ratings for negative emotion questions were generally *higher*—except in the case of "pain"—and the increase was statistically significant for the adjectives "worried" and "bored" (OECD, 2013, p. 87). Similarly, when the order of positive and negative adjectives was varied, Krueger et al. (2009) observed higher ratings of positive emotions in a positive-to-negative order and lower ratings of negative emotions in a negative-to-positive order.

In the life-evaluation context, Deaton's (2012) analysis of data from the Gallup-Healthways Well-Being Index demonstrated the importance of the content of questions (and responses) that precede an assessment of evaluative well-being. This randomized study showed that certain questions about political topics, which apparently altered respondents' feelings while answering the questions, had a substantial impact on evaluative well-being

---

[7] OECD *Guidelines* (2013) includes a more thorough discussion of this literature than is provided here. The OECD report also contains a number of thoughtful recommendations and priorities for future work, which this panel endorses, to improve understanding and deal better with question order and context issues.

as rated using the Cantril ladder. Specifically, "prompting them to think about [politics and politicians] has a very large downward effect on their assessment of their own lives" (Deaton, 2012, p. 19). The magnitude of the effect (about 0.6, or a rung on the Cantril ladder) was comparable to that associated with becoming unemployed. This magnitude of effect translates into a larger impact on average changes because a comparatively small percentage of respondents become unemployed, whereas all can be influenced by question order and context. An assessment of ExWB (global-yesterday adjectives) was placed later in the interview, and the political questions had considerably less impact on those responses. However, it is not clear if it was the "distance" from the political questions or the nature of the ExWB questions that was responsible for small impact. Deaton concluded that these unintended effects linked to context could threaten the internal validity of studies that did not take steps to resolve them.

> RECOMMENDATION 4.2: As part of a general research program to study contextual influences on ExWB measures, survey designers should experiment with randomization of question ordering to create opportunities to study (and eventually minimize) the associated effects. Further work is likewise needed on the effectiveness of buffer and transition questions that precede and follow SWB question modules.

Deaton's analysis is supportive of the notion that including a buffer or transition question between the political questions and the life-evaluation questions would largely eliminate the item-order effect. This was the case when, after an initial period, Gallup added a transition question of the form, "Now thinking about your personal life, are you satisfied with your personal life today?" That this insertion virtually eliminated the item-order effect suggests that careful survey design has the potential to greatly minimize such effects. This finding supports earlier work by Schwarz and Schuman (1997) indicating buffer questions, even a single one, could be effective at reducing context effects. However, they also found that buffer questions that are related to the subsequent SWB questions could prime responses in a way that generates additional context effects. More work is needed to study the frequency with which context and question-order effects arise, their severity, the effectiveness of methods to reduce them, and how they may impact measures of evaluative well-being and ExWB differentially.

> CONCLUSION 4.2: Though not evaluated by the panel in detail, evaluative well-being and even global-yesterday ExWB questions likely benefit from being placed at the front of surveys or, when this is not possible, by the use of buffer questions. Further work is needed on the

ADDITIONAL CONCEPTUAL AND MEASUREMENT ISSUES   83

most effective content and phrasing of these questions. In contrast, for reconstructed activity measures such as the DRM or other time-use formats, respondents need to reinstantiate the prior-day emotional context as much as possible. All SWB questions should appear in the same module of a given survey where possible. Based on the current body of research, the ordering should be questions on evaluative well-being first, questions requiring reinstantiation content next, and ExWB or hedonic questions last.[8]

To summarize the above discussion, much is already known about how to think about these effects. Considerations of context and order are important for deciding how to interpret data as well as how to design surveys. Many questions of design and interpretation can be addressed using fairly straightforward experiments. In many cases, existing research indicates what to do about these biases and what to do, for example, to handle mood effects. Researchers (e.g., Eid and Diener, 2004; Schwarz, 1987) have documented mood changes associated with the weather, question order, or minor events such as finding a dime before answering a question, which in turn influence reported life satisfaction; others have "used structural models to attempt to separate situational variability from random error and basic stability" (Krueger and Schkade, 2008).

4.7 SCALE EFFECTS

Another survey construction issue is the measurement scales used in response formants. At one end, dichotomous scales—for example, yes/no responses—are easy to summarize (as in "x% of this group reported high stress"), so they are useful and understandable. One methodological reason supporting the 0-1 dichotomous option is that it eliminates scale effects (although there is little evidence that they are major). Cultural differences affecting interpretation of terms such as "a lot" are similar to scale effects, and using dichotomous response options may minimize cultural effects; however, there is presently no evidence supporting this contention.[9] The advantage of multipoint scales of the kind favored by

---

[8] This conclusion is consistent with the similar OECD (2013, p. 127) conclusion: "Question order effects can be a significant problem, but one that can largely be managed when it is possible to ask subjective well-being questions before other sensitive survey items, allowing some distance between them. Where this is not possible, introductory text and other questions can also serve to buffer the impact of context."

[9] Extreme responses to scales could represent one form of arousal measurement, discussed in section 4.1. A possible drawback to multipoint scales is that there could be differential group-level reporting patterns associated with certain emotions or sensations—that is, a propensity to choose scores closer to the ends of the scale.

ONS, which uses a 0-10 version, is that they contain much more information. For this reason, the panel agrees with ONS (2011) and OECD (2013) conclusions that a multipoint numeric scale is generally preferably to a dichotomous question structure—though, as always, this hinges on the purpose to which the data will be put.

Going further, for emotion measures, the panel agrees with the following OECD (2013, p. 126) conclusions about how response scales should be labeled and structured:

> there is empirical support for the common practice of using 0-10 point numerical scales, anchored by verbal labels that represent conceptual absolutes (such as *completely satisfied/completely dissatisfied*). On balance, it seems preferable to label scale interval-points (between the anchors) with numerical, rather than verbal, labels, particularly for longer response scales. . . . In the case of affect measures, unipolar scales (i.e., those reflecting a continuous scale focused on only one dimension—such as those anchored from *never/not at all* through to *all the time/completely*) are desirable, as there are advantages to measuring positive and negative affect separately.

Standardized wording, scaling and ordering, question buffering, etc., are all important implementation considerations for which the field does not yet have a full understanding, and for which more research is therefore warranted.

4.8 SURVEY-MODE EFFECTS

Survey mode refers to the vehicle used to ask respondents questions—by personal interview, phone, Internet instrument, and so on. Preliminary results, discussed in this section, indicate that survey mode has a significant impact on responses and, perhaps more importantly, on who responds in the first place. More needs to be known about who is in a given study and who does and does not answer specific kinds of questions (for example, are happier people more likely to respond to the survey?). A big advance would be the ability to gain clear clues about these kinds of selection biases and how to solve them.

Dolan and Kavetsos (2012) investigated the differences between interviewer-administered and telephone-administered responses to the UK Annual Population Survey. The authors examined (a) the impact of survey mode on SWB reports and (b) the determinants of SWB by mode, using the April-September 2011 pre-release of the survey data. Their analysis found large differences by survey mode; in fact, mode effects in the data swamped all other effects. This carries implications for descriptive statistics already published in ONS reports, which ONS has acknowledged (2013, p. 30). This kind of result, similar to Deaton's (2012) findings about question ordering in the Gallup surveys, can seriously undermine a survey enterprise.

The results by Dolan and Kavetsos (2012) are particularly important for cross-region comparisons, because some regions covered by the Annual Population Survey are interviewed via one mode only (the study is based on region W1 respondents only, to avoid self-selection into mode). Their finding was that individuals report higher SWB over the telephone than in face-to-face interviews. Scores for average life satisfaction, happiness, and worthwhileness were about 0.5 points higher in the telephone interviews, and anxiety was about 0.3 points lower. For happiness, the telephone coefficient was three times as large as the (absolute) negative effect associated with being male. That effect is sufficient to offset more than half the effect of widowhood and is more than twice the coefficient of degree-level education; it offsets about a quarter of the effects of unemployment. A large research literature exists on the problem of survey mode effects generally; going forward, it will be crucial to study different survey modalities, including the Internet, for SWB applications specifically.

RECOMMENDATION 4.3: Given the potential magnitude of survey-mode and contextual effects (as shown in findings related to work by ONS and elsewhere), research on the magnitude of these effects and methods for mitigating them should be a priority for statistical agencies during the process of experimentation and testing of new SWB modules.[10]

The OECD *Guidelines* presents a thorough review of the issues and the evidence in the literature, and it offers sensible guidance on next steps:

> Where mixed-mode surveys are unavoidable, it will be important for data comparability to select question and response formats that do not require extensive modifications for presentation in different modalities. Details of the survey mode should be recorded alongside responses, and mode effects across the data should be systematically tested and reported . . . enabling compilation of a more comprehensive inventory of questions known to be robust to mode effects. (OECD, 2013, pp. 127-128)

---

[10] OECD (2013, p. 127) similarly recommends that "details of the survey mode should be recorded alongside responses, and mode effects across the data should be systematically tested and reported."

This section has touched on most, though not all, of the major measurement hurdles facing SWB measurement.[11] Until the issues discussed here are more fully sorted out, using split trial and other experiments, it is hard to make the case for expanding SWB questions into the major U.S. federal surveys.

---

[11]For example, work is needed to better understand and estimate the role of traditionally unobservable characteristics for those who select into a survey (versus those who opt out); innovative methods are needed to ascertain how "happy" people are who refuse to participate in an SWB survey. Related is the effect that being surveyed itself has on other outcomes, bearing in mind that participation in well-being surveys is itself an intervention.

5

Subjective Well-Being and Policy

Informing policy—or at least the potential to do so—is a critical criterion for deciding whether it is worth the time and cost of measuring experienced well-being (ExWB) in national flagship population surveys (for example, the American Community Survey and Current Population Survey in the United States or the UK's Annual Population Survey) or in more focused domain-specific surveys, such as the Health and Retirement Study, the English Longitudinal Survey of Ageing, and various crime, health, or housing and neighborhood surveys. If the relevance or appropriateness of ExWB as an instrument for decision making, policy evaluation, or monitoring purposes cannot be established, then the case for government-supported data collection becomes difficult. An overarching question is whether self-reported ExWB metrics add analytic content above and beyond the existing dashboard of statistics—e.g., those based on income and health data—more traditionally used to measure well-being. In other words, *to what extent do results of subjective well-being (SWB) research go beyond the realm of the interesting and thought provoking (which has already been established) to a point that they might refocus policy or even directly inform it?* This chapter identifies a number of policy areas for which ExWB measures show promise.

A number of recent studies support the validity of SWB concepts and data when applied to policy-relevant social science research. Kahneman and Deaton (2010) and Stevenson and Wolfers (2013) used data collected in the Gallup-Healthways Well-Being Index to estimate the impact of income and income-normalized effects on evaluative well-being (life evaluation) and ExWB; understanding this relationship could be a consideration in tax and social program policies. Oswald and Wu (2009) used data from the Behavioral Risk Factor Surveillance System to rank U.S. states based on hedonic analyses of regional variation in such factors as precipitation, temperature, sunshine, environmental greenness, commuting time, air quality, and local taxes. Diener and Chan (2010) argue that people's emotional states causally affect their health and longevity, concluding that the data are compelling, though "not beyond a reasonable doubt."

From longitudinal prospective studies to experimental mood inductions where physiological outcomes are assessed, the data have shown strong associations indicating that high positive and low negative emotions are likely beneficial to health and longevity. Recent work on well-being and cardiovascular disease finds a comparatively strong relationship between people's emotional states and the behaviors that affect the risk for cardiovascular disease; research on the relationship between evaluative well-being measures and health is more mixed (Boehm and Kubzansky, 2012).[1] A line of research (e.g., Steptoe et al., 2005) has established that SWB measures relate in a predictable manner to physiological measures, such as cortisol levels and resistance to infection.

Developing more robust and comparable measures of people's SWB can also play an important role for decisions aimed at improving the living and working conditions of different population groups, including children or older adults. These measures hold the promise of predicting later outcomes and well-being for children associated with different custodial arrangements or of providing evidence about the relative impact of different factors (such as health status, employment status, transportation and mobility, and social isolation) that prevent older people from living in conditions of greater autonomy. Such measures—many of which should be based on longitudinal data—may shed light on the importance of people's appreciation of their own health (beyond objective measures of their physical functioning) for the quality of various dimensions of their lives. Given the emphasis in the United States, as elsewhere, on enhancing people's physical and mental health—beyond disease prevention—information on SWB, including ExWB, can play an important role in guiding policies and delivering higher-quality services. Another issue is how to measure the effects of painful but needed policies, such as austerity, that produce short-term pain but long-term gain.

The unique policy value of ExWB measures may lie not in assessing how income does or does not relate to an aggregate-level tracking of emo-

---

[1] Similarly, Cohen et al. (2003) examined how Positive Emotional Style predicts resistance to illness. The authors controlled for other social and cognitive factors associated with Positive Emotional Style and compared resistance to rhinovirus or influenza virus of a group characterized by being happy, lively, and calm with a group characterized as anxious, hostile, and depressed. They found significantly different rates of symptom reporting and concluded that Positive Emotional Style may play a more important role in health than previously thought.

tional states but in discovering many actionable relationships that otherwise escape attention: commuting patterns, accessibility of child care, exercise, interaction and connectedness with neighbors or friends, understanding impacts of corruption, presence of neighborhood amenities and other city planning issues, divorce and child custody[2] laws, and the like. Many potential applications rely on analyses of ExWB measures that are tied to time-use and activity data. These targeted areas can be (and have been) improved at many levels, from company policies that improve well-being—and possibly, in turn, improve productivity and lower absenteeism—to community or regional planning policies. ExWB measures seem most relevant and useful for policies that involve weighing costs and benefits when there are nonmarket or not easily quantifiable elements involved—for instance, government consideration of spending to redirect an airport flight path to reduce noise pollution, funding alternative medical care treatments when more is at stake than maximizing life expectancy, or selecting between alternative recreational and other uses of environmental resources.

The possibility of using aggregate-level SWB statistics for broad population monitoring purposes has also been raised. The UK's Office for National Statistics (ONS) has expressed the view that multidimensional measures of the progress of society are needed, focusing on a "triple bottom line": economy, social, and environment and sustainability. ONS states that "overall monitoring of progress" is one possible goal of SWB data. To the extent that this becomes useful, there is certainly consensus (see OECD, 2013; or Office for National Statistics, 2011) that SWB measures should be viewed as one set in the much broader array of indicators through which populations are monitored and policies informed.[3] Statistics capturing trends in a population's health, poverty and income distribution, home production, and environmental degradation are all crucial, as are SWB measures,

---

[2] Child custody and child care discussions raise the issue of whether the SWB of children should be tracked, an issue not addressed in this report. Pediatric SWB measures are being developed as part of the Patient Reported Outcome Measurement Information System, which is designed to produce numeric values indicating patients' state of well-being or suffering and their ability or lack of ability to function. See http://www.nihpromis.org/default [October 2013].

[3] No one is seriously discussing *replacing* other monitoring statistics with an SWB catch-all. The National Income and Product Accounts (NIPA), for example, have proven to be extraordinarily valuable historically, and the core concept is powerful and useful to preserve in something close to its current form. The same can be said of other health, economic, and social "headline" statistics. That said, all measures have limitations and appropriate use constraints. The gross domestic product measure derived from the NIPA is only one piece of evidence among many used for evaluating economic progress and performance. The shortcomings of focusing *only* on market transactions and measuring their impact in terms of market prices have been well documented (e.g., National Research Council, 2005; Stiglitz et al., 2009). Effective social and economic policies require much more.

which may serve to connect the patchwork of social science statistics that, together, create a portrayal of where a people are as a society and where that society is heading. Thus, an SWB "account" or set of indicators would supplement other key social and economic statistics. Sir Gus O'Donnell, chairing a commission on how well-being data can be used by the UK central government, has communicated an urgency in moving toward greater use of SWB for use in policy making—"if you treasure it, measure it." He has outlined a number of policy areas—encouraging altruism and volunteering, community spending decisions, and carbon reduction, to name a few—where he believes SWB data could be used to effect positive changes.

These important goals notwithstanding, the panel does not expect SWB (experienced or evaluative) to produce a single number on the state of the nation or to replace established statistics, such as gross national product (GDP),[4] unemployment rate, or vital statistics. SWB is multidimensional—perhaps more so than measures of market output, unemployment, or mortality rates; there is no comprehensive single measure of happiness or of suffering. ExWB, in particular, does not establish any sort of overall measure of social well-being, but its measurement is proving useful when applied to specific questions, such as evaluating end-of-life care or child custody options.

> **CONCLUSION 5.1: ExWB data are most relevant and valuable for informing specific, targeted policy questions, as opposed to general monitoring purposes. At this time, the panel is skeptical about the usefulness of an aggregate measure intended to track some average of an entire population.**

At this point, evidence about interactions between ExWB and other indicators is inconclusive. For example, on the relationship between income and ExWB, Deaton and Stone (2013b) note that, at least cross-nationally, the relationship between aggregate positive emotions (here, meaning day-to-day ExWB) and per capita GDP is unclear:

> The countries of the former Soviet Union are among the unhappiest in the world, unhappier than the Congo, Benin, or Chad, for example, and Italy and Denmark are unhappier than Mozambique, Sudan, and Rwanda.

They conclude that such revelations cast doubt on using measures of ExWB to provide an overall assessment of human well-being: "While it makes sense for SWB measures to paint a different picture than GDP, it is

---

[4] In reality, the national income accounts and the labor market statistics are also multidimensional. Neither yields a single measure that fully summarizes their rich detail.

hard to credit a measure that says that Denmark is worse off than Rwanda; being happy is a good thing, but other things surely outweigh it in any credible overall assessment of life." It is important to note that Deaton and Stone's point pertains only to data from the Gallup "happiness yesterday" question. As pointed out in the *World Happiness Report* (Helliwell et al., 2012), for life-evaluation questions—namely, the Cantril ladder of life, life satisfaction, and happiness *with life as a whole*—rankings of countries consistently show Denmark near the top and Rwanda near the bottom.

5.1 WHAT DO SWB CONSTRUCTS PREDICT?

An important, but poorly understood aspect of SWB, is its causal associations—both between factors and reported SWB and between SWB and various outcomes. This is, of course, a difficult problem in many areas of social science. Heckman (2000, p. 91) described the difficulty in establishing causal relationships: "Some of the disagreement that arises in interpreting a given body of data is intrinsic to the field of economics because of the conditional nature of causal knowledge. The information in any body of data is usually too weak to eliminate competing causal explanations of the same phenomenon. There is no mechanical algorithm for producing a set of 'assumption free' facts or causal estimates based on those facts."

This critique seems especially pertinent for analyses based on SWB data, given their inherent nature. For pure program evaluation, a full understanding of causality is not always necessary, but in general, we would like to know how self-perceptions of well-being influence behavior, as well as what conditions and factors influence perceptions of well-being. In most analyses, it is not obvious whether positive and negative emotions are the dependent or the independent variables. The link between positive emotions and health appears stronger than the link between negative emotions and health, but we do not know the extent to which high positive ExWB creates better health or the extent to which better health creates conditions for high positive ExWB. Clearly, both can be taking place. The relationships between income and various SWB dimensions could also embody this kind of circular uncertainty.

As described in Chapter 2, experienced and evaluative types of well-being may have very different causal properties, and certain policies may only address one or the other of these dimensions of SWB. Those which aim to enhance longer-term opportunities may even impart negative short-term effects on daily experience. A policy designed to enhance living quality at the end of life, for example, focuses on the hedonic dimension (which is at least one of the objectives of palliative care, that is, relieving suffering), while a policy aimed at enhancing the education and opportunities of youth focuses on life evaluation (and the anticipation of the impact of education). Thinking in terms of process versus outcomes, one can imagine that acquiring the skills and agency to lead the lives associated with high levels of life satisfaction is, at least at times, associated with stress and other experiences that could undermine happiness or even health. One can also imagine respondents with low expectations, agency, or capabilities finding contentment in particular daily experiences, such as socializing and eating, at the expense of longer-term objectives, such as investments in education and health. The example comes to mind of people who are obese and unhappy—but less unhappy than high-obesity cohorts that have even worse health and lower income mobility (Graham, 2008). It is likely that there are comparable effects for smokers, among other examples. In such contexts, considering only one dimension of well-being, such as ExWB in this instance, could lead to bad policy outcomes, and vice versa.

If daily experiences are negative enough, they might overturn the longer-run objectives of policies. A good example comes from George Akerlof's work on identity (Akerlof and Kranton, 2010). He cites work by Robert Foot Whyte on youth in gangs in New York City who receive scholarships to go to top boarding schools. Often they do not fit in at the new schools and find the experience so unpleasant that they drop out. When they return home, they no longer fit into their home environments. The bottom line of the story is that the daily experience eventually determined the long-run outcomes (Akerlof and Kranton, 2010). Krueger and Mueller's (2011) work on the hedonic well-being of the unemployed shows that the longer the sadness associated with failed job searches is prolonged, the more likely they are to quit searching for jobs, ultimately affecting their global life satisfaction evaluations as well.

Momentary feelings and experience drive some health behaviors—eating and smoking habits, for instance—while global memories drive other kinds of behavior, such as economic decision making. For example, a person does not think about his or her car most of the time, even while driving. But, when choosing a car to purchase, the global memory is of the car because the person has been prompted. Thus, ExWB measures can reveal the well-being differences between daily activities better than long-term measures such as life satisfaction (evaluative well-being). This makes the ExWB measures ideal for assessing factors that vary across people's days. In contrast, life satisfaction is more likely to reflect general, long-lasting factors such as unemployment, income, or a happy marriage, although it is easy to see how these circumstances could directly impact ExWB.

ExWB measures may also be capable of uncovering the impact of objective conditions that are themselves not known by individuals. For example, air quality is known to influence mood and behavior, and even life satisfaction (see Luechinger, 2009), but it is difficult for people to recognize such associations and report them, while subjective reports of feelings (which may fluctuate as a function of air quality) may provide more accurate—or at least more useful—information. It is exactly these sorts of associations that the combined use of granular time-use and emotions approaches, such as Ecological Momentary Assessment or the Day Reconstruction Method (DRM), are capable of identifying.

Friendships and socializing (connectedness) stand out as additional factors extremely important to ExWB. Connectedness is also important to life satisfaction and other evaluative well-being metrics, but it may be more important in relative terms to the evaluative well-being of those respondents with less means and opportunity than of those who have greater capabilities and other overarching life objectives.[5] In this instance, agency may be an important mediating factor (Graham, 2011; see also the findings by Diener et al., 2010, on religion and friendships around the world).

Thus, the various types of SWB measures reveal distinctly different things. While people with children tend to evaluate that aspect of their lives as highly important and meaningful, time spent with small children is often reported as the least enjoyable time of the day in time-use surveys (as any busy parent who has had to drop all else at work to take a sick child to the doctor can attest, although the experience hardly results in less love for the child).[6] Understanding this difference—for example, that child rearing can cause quite intensive stress, even in the context of deep affection and it being a desirable aspect of life—could help policy makers better understand the constraints faced by those individuals or cohorts without the means to cope with that stress, among other things.

CONCLUSION 5.2: To make well-informed policy decisions, data are needed on both ExWB and evaluative well-being. Considering only one or the other could lead to a distorted conception of the relationship between SWB and the issues it is capable of informing, a truncated basis for predicting peoples' behavior and choices, and ultimately compromised policy prescriptions.

---

[5] Robert Sampson's Chicago neighborhoods study (Sampson and Graif, 2009) reveals the importance of connectedness to the well-being of neighborhoods. One of many examples is the variation, even among relatively poor areas, in the resilience of different neighborhoods to the 1994 heat wave in the city. Sampson's findings were used to support the creation of a new (for 2013) Neighborhood Social Capital module of the U.S. Department of Housing and Urban Development's American Housing Survey. The survey asks about trust in neighbors, friends in one's neighborhood, interactions, connectedness, and so on. SWB questions might add an insightful dimension to this module, in that they could reveal nonmonetary elements of people's surroundings that influence their well-being.

[6] See, for example, the findings on women in Texas by Kahneman and Krueger (2006). New work by Deaton and Stone (2013a) finds that parents have more positive affect but also more negative affect.

One example where these downsides could occur is considering only ExWB in the case where obese individuals are less unhappy in high-obesity cohorts than in lower-obesity cohorts. A second is considering evaluative well-being when looking at the effects of acquiring skills and agency without considering the stress, function of emotions, and other health effects more closely linked to ExWB.

The evidence suggests that life satisfaction correlates more strongly with external factors such as income and economic region, whereas ExWB measures correlate more strongly with personality. This difference raises questions about how people adapt (discussed in Chapter 4) and about the feasibility of improving ExWB in the long term. A possible implication of adaptation is that, if people are accustomed to living in deplorable economic conditions (so the conditions are no longer reflected in their ratings of pain, stress, and discomfort), their chronic suffering is no less real or in need of policy attention just because they have become used to it. Describing this issue, Sen (1985, p. 14) wrote:

> A person who is ill-fed, undernourished, unsheltered, and ill can still be high up in the scale of happiness or desire fulfillment if he or she has learned to have "realistic" desires and to take pleasures in small mercies . . . the metric of happiness may, therefore, distort the extent of deprivation in a specific, and biased way . . . [and] it would be ethically deeply mistaken to attach a correspondingly small value to the loss of well-being because of this survival strategy.

Deaton, reiterating Sen's point, concluded that:

> we should not base policy on a measure that is subject to hedonic adaptation. Yet the extent to which any particular measure of SWB is actually subject to the adaptation critique is a question that can be investigated empirically, so that it is possible that Sen's concern is hypothetical, or is hypothetical for some measures but real for others. Note also that Sen does not deny the goodness of happiness in and of itself, only that it is an unreliable indicator of overall well-being.[7]

Much of the adaptation question has to do with the distinction between overall life satisfaction and day-to-day experience and with the *time horizon* of interest. Optimization of short-term versus long-term well-being (both at individual and aggregated levels) may imply different policy actions. A program to reduce fat intake or smoking may reduce ExWB in the short run but increase it (via the health covariate) over the long run. Life-

---

[7] Presentation by Angus Deaton to the Panel on Measuring Subjective Well-Being in a Policy-Relevant Framework, December 2012.

cycle modeling and interplay of ExWB measures with evaluative well-being measures will play a role in advancing the assessment of SWB for specific policies.

> CONCLUSION 5.3: The type of ExWB measurement employed for policy use will depend on the specific questions to be addressed. In some cases, global-yesterday measures may suffice, but in other cases a DRM-type measure may be more valuable, because it captures time-use and allows associations between affect and specific activities (which may be selected with the research question in mind). In general, ExWB measures are likely to be most valuable to policy when they (1) capture time-use and (2) associate affect with specific activities, as these kinds of data are amenable to being applied to answer specific questions (as opposed to all-purpose, tracking-type questions).

5.2 WHAT QUESTIONS CAN BE INFORMED BY SWB DATA: EVALUATING THEIR USES

SWB data and statistics are helpful for identifying areas of need and informing policies targeted at subgroups of the population. As emphasized throughout this report, the panel believes the most compelling case for SWB data is its potential to identify populations that are suffering and to help in the study of the sources of that suffering. On the positive-emotion side, there is promising research indicating that health benefits are associated with certain emotional states—but the policy application is less obvious. Here the panel examines several specific possibilities for using SWB data in policy decisions.

5.2.1 The Health Domain

The health domain seems a good starting point for thinking about ExWB and policy. Quality-adjusted life years (QALYs) are usually used to assess health interventions, as they combine the quantity and quality of life into a single metric. Positive ExWB is not normally used in the quality assessments, although negative feelings such as anxiety and depression have been. Because QALYs are usually derived from general population perceptions of how health affects quality of life, they do not capture the actual experience of poor health. While QALYs give a metric for the quantity and quality of life, they have been criticized in several key respects (Garau et al., 2011; Loomes and McKenzie, 1989). As Graham (2008) and others have noted, their adoption to health problems could make policy decisions based on perceptions unreliable and inaccurate. Health policy decision making that utilizes ExWB measures, either in addition to QALYs or incorporated into a revised QALY metric, could be a significant advance. Of relevance to this discussion, it is worth considering the demographic shift in life expectancy and the concurrent increase in the number of older people living with a chronic health condition. A metric based in part on adding years of life may be less useful than a measure of expected ExWB while living with a chronic condition. Interventions could aim to improve ExWB without targeting intractable underlying health problems.

In the United Kingdom, where the burden of social care is increasingly placed on family members, the emotional burden of chronic illness on the family is not adequately captured by the current QALY metric. Given this shortcoming, ExWB could be an appropriate metric for capturing the experience of ill health among patients and their care givers. Similarly, many people with disabilities receive informal family-based care, rather than institutional care. The public and private monetary costs of these two modes differ greatly. A policy-relevant question is, for a subpopulation of persons receiving informal care, are they, net of a range of covariates, experiencing greater SWB than a sample of those receiving formal care? And what is the potential burden, captured in terms of SWB, for the care givers in the informal sector? Might the latter burden be offset to some extent by a higher level of eudaimonia or purpose? These are important but unanswered questions. For evaluating these kinds of policies, simple end-of-day or global-yesterday measures may be sufficient in some cases. For others, the DRM or time-use methods may be more appropriate, although the burden on patients of collecting these data may be too great to merit recommending such methods in all instances.

Richard Frank gave the panel a number of examples from the medical realm for which ExWB metrics are particularly well suited and provide added value.[8] Self-reports of SWB are likely to add useful information in instances where medical interventions have a desired outcome that is something other than merely an increase in life expectancy, where reflections of successful treatment and support extend beyond signs and symptoms and into domains such as functioning and social integration, and where parties other than the patients are affected by treatment and symptoms (care givers, family members, and others).

Valuing end-of-life treatment options is another area that calls out for more nuanced measurement than what simple life-expectancy numbers can provide. Considerable health care costs accrue at the end of life; in many cases, considerable benefits are derived from that care as well. Some agencies, such as the National Institute for Health Care and Excellence (NICE) in the United Kingdom, have explicitly raised the cost-effectiveness

---

[8] Presentation by Richard Frank, Harvard University, to the Panel on Measuring Subjective Well-Being in a Policy-Relevant Framework, March 2012.

thresholds for coverage of end-of-life treatments (by 50 percent in the case of NICE). But is this the "right" policy, and how would one know? To help answer such questions, better data are needed on the impact that end-of-life treatments have on patients and on families and care givers. ExWB is a central part of that impact. Dolan et al. (2013), for example, have assessed the impact of health and life satisfaction on tradeoffs between quality- and length-of-life scenarios. These concepts may be especially important for the end of life, where the balance between predominantly purposeful and pleasurable activities might change (conceivably in either direction).

Terminally ill people often report high levels of purpose, which may translate into a higher reported life satisfaction than many would predict. Cancer patients' will to live has been shown to vary by large amounts over the course of a month, and only somewhat less so over 12-hour periods. These differences can be explained by how the patients felt at the time they were asked about their will to live. Dolan (2008) looked at data on the life satisfaction (evaluative well-being) of cancer patients and found that levels worsen when the cancer is in remission. One possible interpretation of the data is that the imminence of death allows people to "get their house in order" and to solidify a sense of purpose in their lives, whereas remission casts uncertainty in a way that unsettles these thought processes. As with other areas of direct policy applications, more research is needed, including research on the interplay of evaluative well-being and ExWB.

To understand the full costs and benefits of treatment, all of the SWB ripple effects that flow from these circumstances—the immediate effects on patients and their families and the longer-term effects on families after the patient dies—need to be measured and valued. To date, there have been no serious attempts to consider the spillover effects on others over time in such cases. These kinds of results will be of interest to patients deciding upon treatments; clinicians concerned with establishing patient preferences; policy makers deciding on the cost-effectiveness of different interventions; and academic audiences in medical decision making, psychology, and economics.

At key decision nodes or key stages in disease progression, "standard" information could be elicited from patients and close family members on the health-related quality of life according to validated condition-specific and generic measures. Such questions would allow for comparison of the results from that assessment with the results of other studies that have used these measures (including all the recent submissions to NICE). Addition of ExWB measures to such assessments would allow for investigation of the degree to which different people adapt in different ways to their changed circumstances and would enable service providers to reflect more accurately the "epidemiology" of the treatment experience.

5.2.2 Applications Beyond the Health Domain

Many of the policies that may be informed by SWB require data capable of revealing contrasts at the local, or at least subregional, level. ONS is formally looking into possible applications; its case studies include

- Civil Service People Survey—insights into staff well-being to help steer engagement and human resource policies;
- Well-being of job seekers—joining up of mental health and job seeker services;
- Cabinet Office evaluation of the impact of National Citizen Service on the well-being of participants;
- Local government initiatives and policies; and
- Impact of sport and culture on well-being.

Another policy domain where ExWB measures may be useful is in the delivery of benefits. For example, beginning in 2014, the UK government is replacing statements of special educational needs with a simpler assessment process. Parents with a care plan will have the right to a personal budget for their child's education and health support. This policy will enable parents to choose the support and services that they believe are right for their child, instead of local authorities being the sole decision makers. ExWB seems a suitable element to include among the measures used to assess the impact of this change, given the link between autonomy and SWB (although evaluation of a program that aims to affect a child's education and health must surely be centered on the education and health outcomes of those children).

Disability and attendance allowances in the United Kingdom are currently paid to individuals to spend on whatever they wish, to support their independent living. Plans to remove these benefits and place the funds in local authority social-care budgets were shelved after campaigns stressed the importance of personal allowances in people's well-being. But if policy changes such as these were to be implemented, then ExWB could be a suitable complementary measure, along with measures of objective well-being, to assess potential impacts. For evaluating policy changes in the delivery of benefits, simple end-of-day or global-yesterday measures may be adequate, although DRM and time-use assessments may be able to capture specific changes in ExWB while interacting with a child with special needs. It might also be possible to ascertain which activities that disability and attendance allowances support have particularly positive consequences for ExWB. If improvement in ExWB is afforded by access to the social activities and networks that higher disability and attendance allowances would make possible, with consequent improvements in health and reductions in the need for services, then there might be both moral and cost imperatives to giving priority to funding allowances for the disabled and sick.

It is usually assumed that measures of evaluative well-being (life satisfaction) are appropriate for considering work-related policies, but ExWB would add a useful dimension untapped by evaluative well-being measures—such as in the case of policies addressing statutory retirement, unemployment, and working conditions. This extension is consistent with the aforementioned theme that measures of both evaluative well-being and ExWB are needed to provide a comprehensive picture of SWB. For example, in 2011, the policy on retirement in the United Kingdom was changed so that employers are no longer able to force employees to retire at age 65. Being able to continue working if inclined, even if unlikely to change people's overall evaluations of their life, may well increase the positive-emotion aspects of their ExWB. Simple ExWB measures might be adequate for assessing this domain.

The UK policy to increase the statutory retirement age disproportionately affects women in their 50s. For example, a woman currently 55 years of age who thought she would be able to retire at 60 now finds she is not able to receive a state pension until age 66. Women in this age group typically exit the labor force to care for grandchildren, elderly parents, or both, but, without their state pension, they may not be able to afford to leave work to take up these family-care responsibilities. ExWB measures could capture the total burden of paid and unpaid work in late middle age, which other measures of well-being do not capture. Because the United Kingdom and the United States increasingly rely on informal care-giving to support an aging population, it might be important to know more about the decision-making processes involved in balancing paid and unpaid work.

Policies concerned with working conditions, rights, and practices are another domain in which ExWB could play a part. Insecure work has been shown to have almost as great an effect on SWB and the risk for anxiety and depression as does unemployment (Burchell et al., 1999; Ferrie et al., 2002). An interesting and important research question is the extent to which good work conditions and practices improve positive emotions, or at least remove a source of stress. One policy issue such research would obviously inform is whether or not flexible labor market policies are associated with a lower level of positive ExWB in the population. Along these lines, there is a large literature on job satisfaction and the quality of working life, although much of this research has been done in conjunction with overall life satisfaction metrics. Clark et al. (2008) examined the relationships between job satisfaction, wage changes, and future quitting behavior using data from the German Socio-Economic Panel. They found, as did Bertrand and Mullainathan (2001), that job satisfaction was as strong a predictor of the probability of quitting or changing jobs as was wage change. Taylor (2006) investigated day-of-week effects on job satisfaction and SWB. Commuting time and its relationship with ExWB is another often-cited policy application of SWB information (Kahneman and Krueger, 2006). For example, in deciding whether or not to create high-occupancy toll lanes in metropolitan areas, the well-being of people of different incomes who travel the highway has to be examined, along with the network effects—the consequences for those whose travel choices are affected even if they themselves do not use the highway—when estimating full aggregate costs and benefits of a new policy.

Other aspects of planning laws and the built environment could also be evaluated using ExWB measures. Cities that provide easy access to convenient public transportation and to cultural and leisure amenities, that are affordable, and that serve as good places to raise children or to keep older residents better connected have happier residents (Leyden et al., 2011). Data generated by surveys of neighborhood social capital, such as the American Housing Survey (conducted by the U.S. Department of Housing and Urban Development) or Robert Sampson's survey of Chicago neighborhoods (Sampson and Graif, 2009) are useful to researchers investigating whether and how changes to the built environment can promote SWB (the alternative hypothesis being that happier people tend to have more autonomy over where they choose to live). Creating spaces and buildings that encourage and promote SWB, as a worthwhile investment for public health, is an idea that has gained some currency with policy makers. An architectural think-tank book, *Building Happiness: Architecture to Make You Smile*, attests to the acceptance of measuring ExWB to inform policy in this area (Wernick, 2008). Moreover, the relationship between the environment and ExWB can affect policy in areas other than just public health. For example, social and economic benefits go hand in hand with the experiential benefits. Still, more data and more research are needed to better understand what ExWB measures add to assessments of the benefits of green space, transport, or clean and safe urban areas.

Another policy-relevant domain involves the SWB of the unemployed and their experiences as they undergo prolonged job searches, as work by Krueger and Mueller (2011) has shown. Not only did they find that the SWB of the unemployed declines with the duration of unemployment spells; they also found that the time spent involved in job search is particularly unhappy and the unhappiness increases with the time spent in job search (measured both with life-satisfaction and sadness variables). These effects on the unemployed provide an example of how low ExWB related to the process could in the end undermine individuals' incentives to persist, ultimately reducing their capacity to achieve higher levels of evaluative well-being in the future.

Yet another area with possible policy implications is the SWB of refugees or immigrants. SWB metrics can be used to help assess how well they are adapting and assimilating to their new environments, which in turn has repercussions for social stability, investments in children's education, and so on. Does scoring better on either dimension lead to better adaptation or assimilation skills? Do low levels of ExWB, as these people experience the process of adapting to a new environment, lead to lower levels of success in the job market and other areas where greater success would contribute to higher levels of evaluative well-being in the future? Some initial evidence suggests that migrants are more likely to be unhappy prior to migrating, as well as post-migration (Graham and Markowitz, 2011).

Other examples from the literature where ExWB measures have been proposed for informing cost-benefit policy analyses include the following:

- Evaluating trade-offs between inflation and unemployment (Gandelman and Hernández-Murillo, 2009);
- Environmental policies (Ferreira and Moro, 2009);
- Full valuations of cash transfers, earned income tax credit, food stamps, back-to-work programs, and other social policies (Blattman et al., 2013);
- Connectedness (or loneliness/isolation) and health among the elderly; given demographic trends, what will more isolation mean? (Helliwell, 2002);
- Quality dimension of child care arrangements; experiences of parents at work with or without subsidies and/or child care, and with or without health insurance (Brodeur and Connolly, 2012); and
- The effects of different custody arrangements on the SWB of adolescents in divorce situations (e.g., Amato, 1999).

Beyond and apart from informing policies, there is an important role for ExWB measures in advancing research in behavioral sciences, epidemiology, medicine, and even law. Further, such data perform a general information role of interest and value to the public and media. This informing function of basic science often ultimately leads to policy relevance and innovation of science generally. For these reasons, it is worthwhile for governments (and others) to continue to learn about the SWB of the population, especially given that people's well-being, both subjective and objective, is often the ultimate objective of public and private policy. Media and the general public have shown great interest, for example, in information about why some groups—defined by various characteristics or by place—seem happier than others. ONS has explicitly expressed, as part of its Measuring National Well-being Program, the goal of "an accepted and trusted set of National Statistics to help people understand

and monitor national well-being."[9] The underlying belief here is that the need for basic descriptive information is enough justification to warrant data collection, even if causal links between SWB measures and social and economic outcomes (or vice versa) have not yet been established. A broader resonance with the public is driving the recent movements, such as those by ONS and the statistics offices of other countries to implement SWB measurement. At the moment, this informational role is dominated by measures of evaluative well-being. Much of the value of the U.S. Decennial Census—which, granted, is required by the U.S. Constitution for the purpose of drawing political districts and also provides data used for all manner of federal programs—is in its by-product of descriptive information about who Americans are as a society. In general, support of many of the U.S. federal surveys is validated by this extremely important role in producing information regarding the public good.

---

[9] Available: http://www.ons.gov.uk/ons/dcp171766_287415.pdf [October 2013].

6

Data Collection Strategies

6.1 OVERALL APPROACH

Research has shown that it is possible to collect meaningful and reliable data on subjective as well as objective well-being. Subjective well-being encompasses different aspects (cognitive evaluations of one's life, happiness, satisfaction, positive emotions such as joy and pride, and negative emotions such as pain and worry): each of them should be measured separately to derive a more comprehensive appreciation of people's lives. . . . [SWB] should be included in larger-scale surveys undertaken by official statistical offices. (Stiglitz et al., 2009)

The charge to this panel was, in a sense, to deliver an assessment of the extent to which it agrees with the above conclusion of the Commission on the Measurement of Economic Performance and Social Progress, with a primary focus on experienced well-being (ExWB). And, for the most part, the panel does agree: information about the evaluative and experience dimensions of subjective well-being (SWB) is extremely promising for contributing to a fuller understanding of people's behavior and life conditions, and such information should be collected by national statistics offices to the extent that it is practically and financially feasible. However, the panel also recognizes that measurement approaches are not yet fully mature, which generates concerns about their unqualified adoption at this time.

Going further, the panel may appear at odds with the actionable part of the Sarkozy Commission's conclusion that "Despite the persistence of many unresolved issues, these subjective measures provide important information about quality of life. Because of this, the types of questions that have proved their value within small-scale and unofficial surveys should be included in larger-scale surveys undertaken by official statistical offices" (Stiglitz et al., 2009, p. 10). We certainly agree that SWB data have proved their value and are worth pursuing by statistical agencies. The challenge is in interpreting which "larger-scale surveys" are appropriate. In the U.S. context, the bar for getting questions onto the Current Population Survey (CPS) or American Community Survey (ACS) has typically been very high,[1] and it would be advisable for questions to be thoroughly tested and understood before risking using space on those surveys.[2] In contrast, inclusion of SWB questions is warranted in some larger-scale, government-funded academic surveys such as the Health and Retirement Study (HRS) in the United States or the English Longitudinal Survey on Ageing.

The panel also agrees with Stiglitz et al. (2009) that, where feasible, inclusion of SWB questions on the largest population surveys will produce useful information. However, there are many data needs to inform policy in many domains, and it is not obvious yet that the need for SWB data is more critical than the need to include or improve data programs covering other areas where society, economy, and health must be monitored.

A necessary first step is to begin (in the case of the United States) or continue (in the case of the United Kingdom) experimenting with question design and module structure. The UK Office for National Statistics (ONS) is progressing with just such an experimental mode, and it reports remaining open to refining the questions. However, as implied by the OECD *Guidelines* (OECD, 2013, p. 20), there is some commitment to align SWB measurement internationally, and "experimental" status may inhibit rollout and efforts to harmonize. Alternatively, pressure to harmonize may stunt design experimentation and innovation. At this point, the panel sees the risk, associated with greater experimentation, of inhibiting harmonization as the one worth taking for the foreseeable future.

The state of the art has progressed sufficiently far to provide a basis for survey question structure and wording and for experimenting with different

---

[1] Though not always. A civic engagement supplement was added to the November CPS in 2008 and 2010 with somewhat sketchy evidence of how the data would be used and with still-unsettled knowledge of the links between the elements (the module included questions on trust, connectedness, engagement, and other constructs) and social, economic, and health outcomes. Although both social capital and SWB are fertile and important research areas, the panel would argue that the evidence base for structuring SWB data collection is better developed than it is for social capital.

[2] For survey vehicles, such as the CPS and ACS, where space is scarce and highly sought-after, questions typically have only one chance to get things right. Once questions go on and off, they are unlikely to ever be put back on, so the justification for placement must be as solid as possible before the first implementation. Investigation of SWB questions for the major surveys could be a project for the interagency statistics group to consider; a precedent is the process whereby race question were refined by the U.S. Census Bureau—an issue that cuts across surveys.

survey vehicles. These directions can be continued even while uncertainty remains about exactly what kinds of measures will prove most useful to researchers and policy makers and what statistics should be published. In short, this measurement domain is, in the panel's judgment, very much still in the "let a thousand flowers bloom" stage of development.

Evidence about SWB links to behavior, outcomes, and policy levers (causal and otherwise) will continue to accumulate based on research using a range of data sources, public and private. The approach should be to get SWB content on surveys where possible and where it makes sense and to exploit all opportunities to learn more about the data.[3] If, ultimately, a particular SWB metric evolves to a point that warrants "official statistics" status, it may become preferable to have it measured on one official survey and not many; otherwise, different values will be estimated, and the public will be confused. Until then, as a fuller understanding of their properties develops, it is important to try various measures of SWB dimensions and components on a range of surveys. Further, for ExWB specifically—where, relative to measures of evaluative well-being, it is more difficult to envision reporting of some aggregated number—it will be useful to include questions in a range of contexts (time-use surveys, health surveys, housing surveys that include questions about neighborhood amenities and conditions, surveys of the elderly, and other targeted assessments) because it would be beneficial to have information about different sets of covariates for different applications. It is unlikely that an identical module could be simply plugged into different surveys to suit the many envisioned purposes for SWB data; rather, the questions will need to be tailored to the purposes for which a given survey is put. An example is HRS, which includes questions about the connectedness of the elderly to their children and friends, a trait hypothesized to be correlated with happiness and with health outcomes.

6.1.1 The Measurement Ideal

Because SWB has multiple dimensions and its measurement sheds light on people's behavior and life conditions at different levels of aggregation, from the individual up to national and international group comparisons, an ideal data infrastructure would require a multipronged approach.

Large-scale population surveys—such as the four-question module in the UK Integrated Household Survey or the Gallup World Poll—make up one component of a comprehensive measurement program. Data from these surveys, typically drawn from global-yesterday measures of ExWB and

---

[3] At this stage of development, the task of improving measurement methods still lies mainly with academic researchers (many with funding from grant-making government institutions) as opposed to statistical agency staff.

from life-evaluation questions, provide the large sample sizes essential for repeated cross-sectional analyses capable of identifying and tracking suffering or thriving subgroups and for research on special populations such as the unemployed (for whom life expectancy is falling). Such data may prove useful for informing policy at the macro level.[4] The Gallup survey data have also been used for analyses that narrow the focus to specific populations and for city, state, and international comparisons and global hypothesis testing. It is not yet known whether ONS and the Gallup Organization use the right adjectives, or enough adjectives, or if they include the optimal covariates; such an assessment is of course conditional on purposes to which the data will be put. For example, the CPS (in which the American Time Use Survey [ATUS] module resides) is designed to optimize employment measures at specific levels of geographical specificity. Beyond economic contextual needs, there is, for SWB assessments, a need for datasets that include health and geographical covariates.

It is important to establish the right context for a SWB module before attempting to make it permanent. As was learned from the experiences with an official poverty measure, it is difficult to change and improve measurement systems once they become entrenched. Reducing suffering may be an analogous policy goal to reducing poverty, so perhaps lessons can be drawn in terms of how the measurement approach should be developed to best serve that goal. The OECD *Guidelines* (OECD, 2013) will be influential for those countries wanting to go forward with the large population dataset approach, whether this is undertaken by developing new surveys or finding ways to continue ongoing efforts;[5] others will pursue more experimental, smaller-scale approaches before committing—or choosing not to commit. An ever-present consideration for a national statistical office is whether to begin data collection with incomplete knowledge of appropriate structure and its likely value—presumably starting modestly with a small module of questions—or waiting until more is known, when possibly a more expensive, more multidimensional approach can be supported.

The second prong of a comprehensive measurement program is smaller or more specialized data collections. One option is to construct experiments or pilots within existing large survey programs (for example, the ATUS time-use module, which is part of the CPS)—often using outgoing samples that are rotating out of the survey. The American Housing Survey's new

---

[4] The Gallup Organization achieves large sample sizes by surveying people (1,000) every day in its daily U.S. poll. This data collection approach allows things like weekend and holiday effects to be captured. However, for some questions, it may be useful to achieve a similar sample size by surveying much larger groups in a single data collection period but fielding the survey less frequently.

[5] In addition to the United Kingdom, Brazil, Chile, Mexico, and a number of other countries have SWB data collection initiatives up and running.

Neighborhood Social Capital module is another example of a possible home for SWB content; adding ExWB questions to this module would allow researchers to explore links to community characteristics, connectedness, and resilience (an association specifically cited by Stiglitz et al. [2013] as potentially very important). For measuring the role of positive experiences, health surveys provide an increasingly secure foothold, as research strengthens the knowledge base about the links between healthy emotional states and healthy physical states.

The advantage of targeted studies and experimental modules is that they can be tailored to address specific questions of interest to researchers and policy makers—whether about health care, social connectedness of the elderly, city planning, airport noise management, or environmental monitoring. The end objective of these efforts should not be a perfect measure, which does not exist, but measures that generate information that can be usefully combined with other sources in a range of applications. One clear advantage of smaller-scale, specialized surveys is that they can often be supported by funding agencies—such as the National Institute on Aging's support for HRS and the ATUS SWB module—that can ensure the underlying purpose is well thought out.

A third prong to an ideal data infrastructure would consist of panel studies designed to document changes in SWB over time. How individuals' ExWB and evaluative well-being change over time and in reaction to events and life circumstances cannot be fully understood without longitudinal information, which may also help to make progress on causality questions (e.g., does getting married make people happier, or are happier people more likely to get married?). More emphasis should be given to development of longitudinal data sources and within-subject panel design, both to develop optimal measures methodologically and to begin making progress in sorting out causality across measures and events. The policy relevance of monitoring SWB changes over time is clear where, for example, it is important to know the full impact on people of new legislation, such as the Patient Protection and Affordable Care Act, or the full impact on outcomes of experiments such as the Oregon Health Care Study.[6]

Changes in a population's aggregate-level SWB associated with specific events, even very dramatic ones such as the financial collapse or the September 11, 2001, terrorist attacks, can be difficult to detect using broad surveys, even with very large samples (Deaton, 2012). In the case of the 2001 terrorist attacks, Metcalfe et al. (2011) examined consequences for

---

[6]For the Oregon Health Care Study, 6,387 low-income, nonelderly, nondisabled adult participants were selected to be eligible to apply for Medicaid coverage. The comparison group consisted of 5,842 counterparts (with respect to income, age, and disability) who were not eligible for Medicaid coverage.

the well-being of UK citizens using measures of mental distress from the 12-item General Health Questionnaire. They found an impact on happiness roughly equivalent to one-fifth of the average magnitude associated with becoming unemployed.

Anticipated events offer natural experiments and opportunities to test how measures of SWB react to them or to changing conditions. One example is a project to study the SWB impact associated with the 2012 London Olympics. Ongoing research by a team headed by Paul Dolan followed a group in the United Kingdom throughout 2011, 2012, and 2013, using online surveys, supplemented with additional telephone interviews during the 2012 Olympic Games, to track how satisfied subjects were with their life overall, as well as how happy or anxious they were on certain days. The study is also comparing the SWB of people in Paris (which lost the bid to host the 2012 Olympic Games) and Berlin (which did not bid). The motivation for the study is to improve understanding of the impact of big events—for example, is the effect short-lived or is there a longer legacy effect?—in a way that is useful for decision makers considering bringing these types of big events to their localities.

Specialized studies of this kind can often draw supportable inferences from smaller samples than are used in general surveys. Just as panel data have allowed researchers to learn more about the characteristics of poverty (revealing less chronic poverty and more movement in and out of poverty than was once thought), they may be useful for learning about the duration of depression or suffering and whether these conditions are more chronic or if there is extensive movement by individuals in and out of groups defined by these states. It is difficult to study these phenomena without panel data that are collected fairly frequently. Schuller et al. (2012) reviewed the contribution of longitudinal data in analyzing SWB responses for a range of key well-being domains, such as relationships, health, and personal finance. A panel structure also creates pitfalls. For example, asking a panel of individuals SWB questions on, say, a quarterly basis might give rise to a focusing effect, where the first (or previous) response acts as a reference point for subsequent ones because individuals might recall their previous response (Dolan and Metcalfe, 2010).

A final prong of an ideal data collection is real-time experience sampling. As described in Chapter 3, momentary sampling methods have been central to SWB research but largely out of practical reach for adoption by national statistical offices. However, rapid changes in technology and in the way the public exchanges information have brought the world to a point where momentary assessment techniques may now be on the horizon for national statistics. Precisely knowing how people are doing emotionally and what they are doing in the moment can shed light on the effects of commuting, air pollution, child care, and a long list of areas with clear ties to policy. As the ways in which government agencies administer surveys change (with response rates declining and survey costs continuing upward for conventional data collection methods) and as monitoring technologies continue to rapidly evolve, new measurement opportunities will arise. Considered in terms of comparative respondent burden, it may soon become less intrusive to request a response to a modern electronic Ecological Momentary Assessment (EMA) device, or perhaps a smartphone beep, than to ask respondents to fill out a long-form survey. So, while EMA may not be practical for the ACS or CPS for the foreseeable future, real-time analyses may be (or become) practical for a number of other surveys, particularly in the health realm.

Which elements to pursue of the four-pronged approach outlined above is a matter of national and specific program priorities. ONS is moving forward on a broad survey of SWB measures, and it may later begin adding granular time-use and targeted survey components. The plan would be quite different for the United States, where ATUS has been temporarily in place as an "experimental module" but where a SWB module for the largest population surveys is lacking. In the meantime, it is a boon to researchers that different organizations are taking the lead in different areas so that the relative merits of different approaches can be assessed.

6.1.2 Next Steps and Practical Considerations

Appropriate next steps for the statistical agencies will be dictated to a large extent by perceptions of the state of maturity in the evolution of SWB measures. Evaluative well-being has for some time been measured with one or two questions in many large-scale surveys, and that approach can and will continue to be applied at a relatively low cost in national and international surveys (as done now by ONS or by the Gallup Organization). ExWB is less well understood and less well tested—though, certainly, some questions still remain about evaluative well-being as well—and therefore its measurement is more challenging from a survey methodology standpoint.

Given that detailed time-use surveys are expensive and burdensome to conduct and that simple ExWB measures (e.g., global-yesterday measures) used in larger-scale surveys such as the Gallup World Poll seem to track well with the more detailed measures, one approach would be to include simple measures of ExWB in a set of large-scale surveys. The results could then be rounded out using more detailed surveys of time use, such as ATUS, on a more targeted, small-scale basis. As acceptance and validity become more established, a more aggressive move to add content to U.S. federal surveys can be supported. At this point, however, more research and testing are needed before the federal statistical system should settle on a specific approach or create an "official series" for ExWB comparable to, say, the unemployment rate.[7] There simply is not enough known yet about ExWB over time to present it in an official government series.

> CONCLUSION 6.1: SWB is an exciting and potentially very important construct that adds content to and could influence the direction of policy debates. Recent research has rapidly advanced our understanding of the properties of ExWB measures and their determinants. This promise notwithstanding, more research and assessment are needed before ExWB is included as a regular and permanent component on flagship U.S. surveys, such as the ACS and CPS. ExWB metrics are not yet ready to be published and presented as "official statistics."

Although the level of confidence needed for an official series (which becomes less methodologically flexible than satellite or experimental data) has not yet been established, SWB modules, including questions on both ExWB and evaluative well-being, are appropriate for inclusion in more targeted surveys, such as the ATUS or those administered by various health statistics agencies. The issues described in this report can likely be resolved (or better understood) through experimental pilots and targeted surveys and from further study of results from ONS, the Gallup Organization, ATUS, and other current activities.

> RECOMMENDATION 6.1: ExWB measurement should, at this point, still be pursued in experimental survey modules. The panel encourages inclusion of ExWB questions in a wide range of surveys so that the properties of data generated by them can be studied further; at this time, ExWB questions should only be considered for inclusion in flagship surveys on a piloted basis. Numerous unresolved methodological issues, such as mode and question-order effects, question wording, and interpretation of response biases need to be better understood before a module should be considered for implementation on a permanent basis.

More of the research recommended in this report should be completed (not all by statistics agencies) before committing to a particular version for national time series.

The above statements raise the difficult question of what the criteria are for establishing the level of confidence needed for an official series. As described in Box 6-1, national and international statistical offices take some care to define what official statistics are and to specify the roles that they serve. Assessments of reliability, accuracy, data interpretation, proven policy

---

[7]The panel notes that there are still multiple series for unemployment measures, and there is still methodological debate about them.

> **BOX 6-1**
> **Fundamental Principles of Official Statistics,
> from the United Nations Statistics Division**
>
> **Principle 1. Relevance, Impartiality, and Equal Access**
>
> *Official statistics provide an indispensable element in the information system of a democratic society, serving the government, the economy and the public with data about the economic, demographic, social and environmental situation. To this end, official statistics that meet the test of practical utility are to be compiled and made available on an impartial basis by official statistical agencies to honor citizens' entitlement to public information.*
>
> There are many elements to this principle. First, official statistics are one of the cornerstones of good government and public confidence in good government. Official statistics, by definition, are produced by government agencies and can inform debate and decision making both by governments and by the wider community. Objective, reliable and accessible official statistics give people and organizations, nationally and internationally, confidence in the integrity of government and public decision making on the economic, social and environmental situation within a country. They should therefore meet the needs of a range of users and be made widely available.
>
> Second, to meet the test of practical utility, statistics must be relevant, of a quality suitable for the use made, and in a form that facilitates easy and correct use. The key to achieving this is maintaining an understanding of what statistical information users want and how they want it.
>
> SOURCE: United Nations Statistics Division, see http://unstats.un.org/unsd/goodprac/bpaboutpr.asp?Recld=1 [September 2013].

relevance, and credibility among data users certainly figure into the decision to establish an official series, but it is essentially an iterative process whereby data are first deemed worth collecting, then used to produce pilot or test statistics, and sometimes rising to be published as an official series (as in the case of the consumer price index or unemployment rate).[8] There is no objective, bottom-line criterion indicating when statistics become qualified to be an official series. However, the criteria listed above are part of building a strong case for taxpayer support and for the potential sustainability of a measure.[9]

---

[8] See *Principles and Practices for a Federal Statistical Agency: Fifth Edition* (National Research Council, 2013), a report periodically updated by the Committee on National Statistics, for a thorough discussion of these issues.

[9] For example, a "civic engagement" module was added as a supplement to the U.S. CPS in 2008 and 2010. There was some political and researcher support for the module, but the

In thinking about plans for the United States specifically, an important distinction is that between an official statistical series and government data collection more generally.[10] In the view of this panel, the concept of ExWB is certainly ready for the latter but not yet the former. In the meantime, as data-driven research results accumulate in ways that support (or do not) official SWB statistics, there are actions that the statistical agencies can take to help move things forward.

There is, of course, a danger in being too timid with recommendations for moving into new measurement areas. After all, how can research and development occur without data creation and without risk? In this spirit, Conclusion 6.1 and Recommendation 6.1 should not be interpreted as a knock against the ambitious work undertaken, and the progress being made, by ONS. Many of the fixed resources—both intellectual and financial—for adding the four-question module to the Integrated Household Survey (and others) have already been expended, and the results to date have created an excellent opportunity to begin analyzing data properties, interpreting the results, and generally using them as a test bed for further development of SWB measurement. As researchers take advantage of this emerging data source, much may be learned about the SWB of the UK population and about next steps in developing effective and useful SWB modules. Further, ONS has stated the view that "National Statistics" status does not preclude further refinement.[11]

6.2 HOW TO LEVERAGE AND COORDINATE EXISTING DATA SOURCES

Although researchers have benefited enormously from data collection by the Gallup World and Daily Polls, the World Values Survey, and others, there is clear value (complementary at the very least) in anchoring data collection work in government statistical systems. Government surveys often

---

supplement was dropped for 2012, perhaps in part because the case for its continuation had not been made clearly enough by these criteria. In contrast, the research support and use for the ATUS module of the CPS has been quite broad, and the case for its continuation has been easier to make. The bar for an official series (say of time-use patterns or civic engagement) would be much higher still.

[10] Although it is not within the scope of the panel's Statement of Task, the panel cannot ignore the current political and budget climate, which makes the practical hurdles to introducing new survey content quite high. Furthermore, the decentralized nature of the U.S. statistical system creates additional complications for launching a well-coordinated effort analogous to what ONS has done for the United Kingdom.

[11] For statements on the experimental nature of the ONS data collection initiative, see: http://www.ons.gov.uk/ons/rel/wellbeing/measuring-subjective-wellbeing-in-the-uk/first-annual-ons-experimental-subjective-well-being-results/first-ons-annual-experimental-subjective-well-being-results.html [October 2013].

include rich sets of covariates, large sample sizes, and comparatively high response rates, along with the potential to link with administrative and other data sources. Dwindling budgets may eventually call some of these advantages into question, yet government surveys are stable and likely to be an important public good for many years to come.

To be realistic, agitating for entirely new, large surveys of SWB seems unlikely to pay off in the foreseeable future. In the current budget climate, the statistical agencies will have to be more opportunistic, which likely means formulating a strategy to embed SWB question in existing instruments. ONS has already followed this course, adding modules to the Annual Population Survey in 2012 and the Opinions Survey in 2011. The strategy described above is not too different from that taken by ONS, which is mixing large and small survey instruments. In the United States, a comparable strategy would be to add modules to the ACS or CPS—but the panel has already discussed the practical difficulties of doing that. Nevertheless, beyond these broad population surveys, options do exist to add questions to more targeted instruments.

6.2.1 SWB in Health and Other Special-Purpose Surveys

Several surveys provide a platform for SWB measurement in health domains. One of these, the HRS, is a nationally representative longitudinal survey of more than 26,000 Americans over the age of 50. Conducted every 2 years, HRS has included satisfaction-of-life questions as well as ExWB questions (there was a hedonic well-being module as recently as 2012). The funding agencies—the National Institute on Aging and the Social Security Administration—determined that these questions were useful for generating insights into the health and work transitions of older Americans. HRS, which is conducted by the University of Michigan, includes a wealth of contextual information—on demographics, income, wealth, employment status, health, and disability—making it all the more attractive as a home for SWB research. It is an excellent example of how inclusion of SWB in a more targeted way can lead to rich investigation of well-defined questions, such as how disability in older populations relates to their emotional states (Daly and Gardiner, in press) or how their health is affected by family connectedness and support (National Research Council, 2010). The English Longitudinal Study of Ageing creates similar research opportunities for UK studies. The National Longitudinal Study on Youth, conducted by the U.S. Bureau of Labor Statistics, is another survey that would be useful for studying ExWB alongside work and other factors at the younger end of the age spectrum.

While these kinds of longitudinal datasets are extremely useful for studying impacts that accompany changing life circumstances, repeated cross-sections are sometimes needed to provide information about the evolution of the population. There are also sample-size trade-offs. The questions of interest will dictate whether it is preferable to survey more respondents cross-sectionally or to survey fewer over multiple periods. For national (aggregate-level) statistics, large samples are needed to pick up change so that subpopulations affected by events (e.g., the unemployed; people in New Orleans post-Katrina) can be captured. Krueger and Schkade (2008) provided a straightforward assessment of how to assess the reliability of some SWB measures. A related issue is whether to field surveys intermittently (as in employment surveys) or continuously (as in the Gallup Daily Poll). Other than resource constraints, there seems little reason against a continuous mode.

Among other strong candidates for SWB data collection in the U.S. statistical apparatus are the National Health Interview Survey and the National Health and Nutrition Examination Survey. These surveys include the necessary covariates to study the influence of health and health care on SWB (and possibly vice versa). For these surveys, there is a clear policy rationale related to health care delivery (for patients and care givers) for embedding SWB questions. Similarly, the Behavioral Risk Factor Surveillance System—a repeated cross-sectional survey, which includes a county-level identifier and questions about SWB—has been used to study life-style choices and SWB (focusing mainly on evaluative well-being). For example, Brodeur (2012) examined the impact of smoking ban policies (at the county level) on self-reported life satisfaction, using the Behavioral Risk Factor Surveillance System and the Needham Life Style Survey, both of which include a broad set of variables such as household income and smoking behavior.

The Survey of Income and Program Participation (SIPP), which has historically reflected an interest in self-assessments of well-being, represents another option. SIPP is a natural fit for SWB measurement because of the wealth of income and program-activity questions that form the core of the survey. Kominski and Short (1996) noted the relationship of income to SWB versus other factors and recognized the possibility that some members of a population may have objectively low levels of income (and commodities that can be purchased with that income), yet still be relatively "well off" if other aspects of their lives act to compensate in some way. As an example, they noted that an extensive social support system had been shown in other research (e.g., Helliwell and Putnam, 2004) to significantly offset some of the disadvantages associated with low income and low wealth. Developers of the SIPP recognized that these kinds of research questions require subjective self-assessments of one's quality of life.

A 1978 pilot module of SIPP did ask about respondents' normative self-assessments, using a seven-point "delighted-to-terrible" scale to describe their "life as a whole."[12] Later, during the early 1990s, an interagency group of researchers considered how to develop a set of questions that could be added to the SIPP to elicit a broader (than just income) concept of well-being from survey respondents. The working group was tasked with developing an "extended well-being" topical module for inclusion on the 1991 and 1992 panels of the survey (Kominski and Short, 1996). This module did not ask now-conventional questions for measures of evaluative well-being or ExWB but instead asked about various domains: housing conditions, crime conditions, neighborhood conditions, presence of help when in need, food adequacy, etc. The point here is that the SIPP is an appropriate context for targeted research questions about benefits trade-offs that could be fruitfully supplemented with SWB information. The policy application here is quite clear: How to value the bundle of "goods" (including nonmarket goods and services) provided to low-income families.

Another candidate for ExWB data collection is the large American Housing Survey (AHS), which is overseen by the Department of Housing and Urban Development and conducted by the Census Bureau.[13] Inspired by the research of Robert Sampson on Chicago neighborhoods, the survey will include a new module in 2013 called the "Neighborhood Social Capital Module," which was created as a "rotating topical module that collects data on shared expectations for social control, social cohesion, and trust within neighborhoods, and neighborhood organizational involvement."[14] The AHS survey is conducted with a large, geographically diverse sample, which will enable detailed neighborhood social-capital assessments to be produced for 25 metropolitan areas. Adding SWB questions to the AHS would allow researchers to explore the relationship of SWB measures with community characteristics (the magnitude of income disparities, provision of social services, etc.). Social context is an association that has been studied in some detail by Helliwell and Putnam (2004), among others; it is cited specifically by Stiglitz and colleagues (2009) as central to population well-being.

The Panel Study of Income Dynamics is another option for ExWB questions; it would be particularly useful for researchers studying the relationships between care-giving arrangements, connectedness, health, and SWB. It offers a large, representative national sample of U.S. households and uses subsets of respondents assessed in multiple waves. The 2001 and 2003

---

[12] The seven-point scale, together with the first two items, was originally developed and extensively tested by Andrews and Withey (1976).

[13] One attractive feature of this survey is that it is quite large; 179,000 responses are expected, which is substantially larger than the CPS supplements, and it is longitudinal.

[14] This text is from the Office of Management and Budget supporting statement for this data collection initiative, which can be found at http://www.reginfo.gov/public/do/DownloadDocument?documentID=369083&version=0 [October 2013].

waves included items on SWB. Also, Child Development Supplements that have included evaluative well-being questions about how often during the past month respondents had felt "(1) happy, (2) interested in life, and (3) satisfied"—have been attached to past waves of the PSID.

> RECOMMENDATION 6.2: ExWB questions or modules should be included (or should continue to be included) in surveys where a strong case for subject-matter relevance can be made—those used to address targeted questions where SWB links have been well researched and where plausible associations to important outcomes can be tested. Good candidates include the Survey of Income and Program Participation (which offers income, program participation, and care-giver links); the Health and Retirement Study (health, aging, and work transition links); the American Housing Survey's Neighborhood Social Capital module (community amenities and social connectedness links); the Panel Study of Income Dynamics (care-giving arrangements, connectedness, and health links); the National Longitudinal Survey of Youth (understanding patterns of obesity); and the National Health Interview Survey and the National Health and Nutrition Examination Survey (health and health care links).

If harmonized modules were developed that were short enough, they could in principle be included in a range of surveys. However, for surveys with a specific orientation (e.g., understanding the conditions of retirees or the time use of individuals) it would typically be preferable to tailor questions to research objectives. One possible benefit of an initiative to design a standard ExWB module or instrument (perhaps developed by a research network, the National Bureau of Economic Research, the Russell Sage Foundation, the Roybal network, or through a pilot study competition) is that it would encourage discussion of where the measures are useful and where not, and it may help to reframe the discussion about what are the clearest policy applications.

Also, if inclusion of a uniform ExWB question or module into a number of surveys were considered, global-yesterday measures would be the likely default instruments, as they are short by design and flexible in terms of survey mode (i.e., the time of day when the question can be asked). However, they are more limited in the scope of detail that can be collected relative to something like the ATUS SWB module, or certainly the EMA-type methods.

6.2.2 Taking Advantage of ATUS

The ATUS SWB module is, at this time, the most important U.S. government data collection on ExWB. Funded by the National Institute on

Aging and overseen by the Bureau of Labor Statistics, an SWB module was included in ATUS in 2010 and 2012, but there are no plans currently in place to field it in 2014 or beyond. The ATUS SWB module is the only U.S. federal government data source of its kind—linking self-reported information on individuals' ExWB to their activities and time use. As described in Chapter 3, time-use data derived from the Day Reconstruction Method (DRM) or a modified DRM is essential for linking ExWB to activities and, in turn, to policy levers. The fact that ATUS itself is a supplemental module to the CPS, which is focused on labor market and other economic information, adds further value. Research (e.g., Krueger and Mueller, 2008, 2011) has shown that long-term unemployment is strongly linked to suffering, so this relationship can potentially be studied.[15]

Time-use surveys are needed to determine how people change their time allocations and to indicate which activities are most enjoyable and which are most miserable.[16] Questions about "overall happiness yesterday" miss much of what is interesting in this regard. Important research has already been conducted using the time-use data (for example, that cited above on the effects of unemployment and job search on people's SWB). If attaching SWB questions to an existing instrument can be done at low marginal cost, it seems a good value (see Appendix B for the panel's interim report on the ATUS SWB module). Work conducted with ATUS—sometimes in combination with other data sources—has indicated the potential of the module to contribute to knowledge that could inform policies in such areas as health care and transportation. If a policy changes time use—typically the most valuable market and nonmarket resource in an economy—then it is easy to make the case for data collection.

CONCLUSION 6.2: Time-use data are being collected by the U.S. government, and self-reported well-being questions add an important dimension to such data. The ATUS SWB module is practical, stable, inexpensive, and worth continuing as a component of ATUS. Not only does the ATUS SWB module support research; it also generates information to help refine SWB measures that may be considered for future additions to official statistics.

---

[15] Effective January 2011, the CPS was modified to allow respondents to report durations of unemployment up to 5 years. Prior to that date, the survey allowed reporting of unemployment durations of up to only 2 years; any response greater than 2 years was entered as 2 years.

[16] Not all time-use policy questions require ExWB information. Increased time spent in commuting is known to have a negative impact on people's emotional states; one only needs to look at activity-based average scores and allocation of time to compare one state and another in a general way.

Extending the material developed during the course of the panel's early deliberations and presented in its interim report on the ATUS SWB module (see Appendix B) are the following additional conclusions:

- *Continuation of the ATUS SWB module enlarges samples by allowing pooling of data across years.* This enables more detailed study and comparison than has been possible to date of population subgroups, such as people in a given region and specific demographic groups (e.g., young people, the elderly). Because two new questions—one on overall life satisfaction and one on whether respondents' reported emotional experiences yesterday were "typical"—were introduced to the module in 2012, additional waves of the survey will allow assessment of changes in response to those questions over time (although the responses over time will not be from the same respondents).
- *Cost and other effects on ATUS.* As a supplement to an existing survey, the marginal cost of the SWB module, which adds about 5 minutes to ATUS, is small. While further study of the module's effects on response and bias in the main ATUS should be undertaken, it appears likely that these effects are modest because the module comes at the end of the survey, after respondents have already been asked to report their activities for the preceding day.
- The ATUS SWB module could be the basis for a standardized set of questions that could be added to other time-use surveys around the world, which together might provide useful comparative information across different populations.

ATUS provides an appropriate vehicle for experiments to improve the structure of abbreviated DRM-type surveys. Experimental modifications to consider include

- *Split sample surveys.* Half of ATUS respondents could receive one question while the other half gets another; this would be useful for testing such things as sensitivity to different scales and question wording.[17]
- *Finding the optimal number of activities on which to collect ExWB information.* It is not obvious that three activities is the optimal number of activities to include on the ATUS SWB module. It may be useful to ask about ExWB associated with more activities in order to increase the reliability of daily estimates. Importantly,

---

[17] In its well-being survey, ONS has used, or plans to use, split trials to test such things as sensitivity to different scales, question wording, and order and placement of questions.

sampling more episodes increases the power to examine activity-specific effects, which may be particularly valuable for addressing policy questions. Doubling or even tripling the number of episodes may be cost-effective, although that benefit would have to be weighed against considerations of participant burden and the potential impact on response rates.

- The data characteristics that emerge from sampling three consecutive activities, in which the first in the sequence is randomly selected, could be tested for comparison with the current structure, in which all three are chosen randomly. Questions of interest include what additional things could be learned (e.g., how emotional impact of one event may carry over to others) and what would be lost from such a question structure.
- *Selecting the "right" positive and negative emotion adjectives for module questions.* As described in section 2.1, research supports the separation of positive and negative states but, more generally, should the SWB module be focused more on suffering or happiness? The module could experiment with different adjectives and how interpretation varies across populations.
- *Additional or replacement questions for consideration.* A possible example is adding a question or two about sleep, such as: "How many hours of sleep do you usually get during the week?" or "How many hours of sleep do you usually get on weekends?" The objective of such questions would be to find out if respondents' reports about behaviors and emotions—feeling happy, tired, stressed, sad, pain—are influenced by (chronic) sleep deprivation or other sleep patterns.[18] A methodological question is how well people recall the previous night's sleep.
- *Selecting among competing measures of evaluative well-being.* Is the current Cantril approach, which is perhaps the most remote from ExWB measures, optimal? Alternative versions of the evaluative well-being measure are common in the literature.

---

[18] This idea was raised by Mathias Basner, of the University of Pennsylvania School of Medicine, who noted that self-assessments of habitual sleep time overestimate physiological sleep time and that estimates of habitual sleep time based on ATUS overestimate self-assessments of habitual sleep times found in other population studies. Therefore, he suggested, it would be very elucidating to compare self-assessments of sleep time for the two questions suggested above against estimates based on ATUS responses for the day before the interview day (public comments for the ATUS SWB module: see http://www.reginfo.gov/public/do/DownloadDocument?documentID=120293&version=0 [October 2013]).

6.3 RESEARCH AND EXPERIMENTATION— THE ROLE OF SMALLER-SCALE STUDIES, NONSURVEY DATA, AND NEW TECHNOLOGIES

At this point in the conceptual development of SWB measures—and in keeping with the panel's conclusion that it is not yet useful to construct a national measure for general monitoring purposes—data collection should be carried out using targeted or specialized tools and in experimental modules of existing surveys. Smaller-scale studies have already shown their potential to inform the development of survey measures of SWB and to be used in substantive research applications. An example is the Krueger and Mueller (2012) study of how job search affects dimensions of SWB among the unemployed using a repeated survey of 6,025 unemployed workers in New Jersey.

> **RECOMMENDATION 6.3:** For ExWB, the data collection strategy of the statistical agencies should remain experimental until data properties and correlative and causal relationships among variables are better understood. This means more research and preliminary testing before committing to particular approaches (e.g., to a given survey module structure).

Beyond the statistical agencies, it is likely that researchers will increasingly exploit alternative, nontraditional survey sources to learn more about SWB. One example is the study of the SWB impact associated with the London Olympics using multiple survey modes (including Internet), cited in section 6.1.1. Social media data and other kinds of organic data (those, such as administrative records or company-maintained information, produced initially as a by-product of nonstatistical purposes) may become increasingly useful for shedding light on trends in people's emotional states. Word-mining exercises have been used to show patterns in emotional states—for example, a Facebook happiness index showed the standard weekend and holiday effects and expected changes associated with major events, such as disasters. Additionally, analyses of data generated by social media and other Internet activities will produce insights relevant to public policy beyond those focusing primarily on aspects of negative experience such as distress or pain. As illuminated by social or political movements such as the Arab Spring and by mass protests across the world ranging from anticapitalist movements to demonstrations concerning police behavior or health reforms, other negative feelings such as collective anger and sense of injustice may be as important in the public policy context as individual experiences of distress. Not much is known about these collective experiences, and the tools have not yet been developed for studying them carefully, but they are surely important; measuring and understanding them would be a significant benefit to public policy. Big data will play a role in this research.[19]

The Mappiness project (mappiness.org.uk), designed to investigate well-being effects to the public associated with open green space in the London area, delivers instant feedback on how Mappiness app-holders feel: happy, relaxed, and awake. It allows monitors to look at individual-level variation for people located in different outdoor environments. Potential applications envisioned by the creators include assessing interventions in the form of random controlled trials of such things as remediation or green exercise, assessing impacts of events such as the Olympics, and natural and recreation resource monitoring. This project provides a clear example of the emerging methods to capture SWB in the context of EMA measures and the role of portable recording—in this case the use of cellphones and global positioning system (GPS) tracking. New measurement techniques such as the geospatial cellphone responses in the Mappiness project are now making it possible to consider EMA-type data in survey contexts. The British Millenium Cohort Study is considering use of geospatial cellphone responses as a post-survey supplement.

There are still major unresolved data quality and representativeness issues in this world of new data and big data. For instance, the sampling properties are largely unknown for data generated by social media, phone records, Internet usage, and the like. A bright red flag of caution needs to be attached to these data sources, acknowledging the unknown distributional characteristics of various underlying subpopulations. This is sure to be a major emerging statistical research topic. Social media data need to be scaled, and the best methods are likely to change as the penetration of various media and technologies evolve. One must also be careful not to clump all kinds of new technologies or big data together. For instance, a Facebook index may not work well for objective statistical analysis, but an iPhone bleep test of a carefully sampled population might—or vice versa for some questions. In the Mappiness project, within-individual confounding is possible; that is, causal pathways may run in both directions: people may

---

[19] While avoiding a formal definition, Capps and Wright (2013) usefully contrast official statistics and big data in terms of database size, dissemination timing, nature of data use permission practices, costs of production, and data collection design. Sources of big data cited by the authors include "data that arise from the administration of a program, be it governmental or not (e.g., electronic medical records, hospital visits, insurance records, bank records, and food banks); commercial or transactional digital data . . . (e.g., credit card transactions, online transactions; sensor data (e.g., satellite imaging, road sensors, and climate sensors); GPS tracking devices (e.g., tracking data from mobile telephones); behavioral data (e.g., online searches about a product, service, or any other type of information and online page views); [and] opinion data (e.g., comments on social media)."

go to certain places when they are happy, or they may be happy because they are at that location.[20]

> CONCLUSION 6.3: For now and the immediate future, the primary means for measuring and tracking ExWB, and SWB more broadly, remains population surveys. Neither the practical and economic challenges to "traditional" survey methods nor the promise of alternative ways for measuring the public's behaviors and views have reached a point where it is sensible to transition away completely from the former.

Thus, the panel agrees with the view expressed in the report of the Commission on the Measurement of Economic Performance and Social Progress that "reliable indicators can only be constructed through survey data" (Stiglitz et al., 2009, p. 184). However, this constraint is likely to change going forward, partly out of practical considerations concerning the cost and viability of conducting large government surveys. Survey research is facing numerous challenges involving the impact on survey response rates of both technological factors (answering machines, mobile phones, etc.) and sociopolitical developments (respondent "burnout" from the proliferation of polls, mistrust of polls and pollsters, etc.). Lower response rates in turn affect the reliability and validity of telephone-implemented survey findings. Surveys such as the CPS (conducted through a combination of in-person and telephone instruments) that have maintained very high response rates (92-94 percent for the core CPS in 2003-2005) are extremely expensive to conduct. Their cost raises concerns about their sustainability and creates a high-stakes competition for the limited space available on their questionnaires.

Partly in response to these pressures, online surveys have emerged, some with promising results. Often, results from these surveys are of value not because they provide valid population-level information (though some panels are working to achieve this goal) but because they may offer a good laboratory for testing different approaches and hypotheses before embarking on larger, more expensive, and more burdensome programs. They may, for example, offer opportunities to study mode effects or to test different adjectives describing emotion, experience, or life satisfactions. More broadly, the emergence of big data (which consists mainly of data generated for purposes quite different from those driving government surveys) that can be captured from a variety of (largely though not exclusively) digital

---

[20]The Mappiness developers note that between-individual confounding should not be a factor because their model is estimated exclusively from within-individual variation (MacKerron and Mourato, 2013).

information and communication technologies, coupled with advances in computational science analytic techniques, raises the possibility of developing less-obtrusive indicators of citizens' well-being, behaviors, and opinions. Researchers have, for example, accessed Twitter to study word use associated with different circumstances such as job search (Antenucci et al., 2012). Such exercises can be used to study changing word use in the population in order to better understand how respondents communicate.

Researchers working at the University of Pennsylvania Computer Science Department have begun conducting research based on the idea that:

> The words people use on social media such as Twitter, Facebook, and Google search queries are a rich, if imperfect, source of information about their personality and psychological state. [They] are developing methods to estimate variation in subjective well-being over time and space from social media word use . . . [and] studying the variation in use of words relating to PERMA (Positive emotion, engagement, relationships, meaning, and accomplishment), and how these correlate with Gallup poll answers and CDC data at the State level.[21]

Similarly, Quericia et al. (2012)—also working from a computer science background—engaged in a project to track "gross community happiness" for physical communities (London, in this case) from tweets. To this end, they examine, for a number of communities, the relationship between sentiment expressed in tweets and community socioeconomic well-being. They "find that the two are highly correlated: the higher the normalized sentiment score of a community's tweets, the higher the community's socioeconomic well-being" (p. 265).

Companies such as Knowledge Networks[22] have made strides in online research, and online surveys are increasingly common in academic scholarship. However, questions remain regarding their ability to fully substitute for more traditional survey modes, and more independent comparative research is needed. Nonetheless, it is important that government agencies follow these developments so they are prepared to adjust the ways they gauge SWB and other important measures in the future. This will entail monitoring survey data collected by private and other public organizations in order to assess needs, determine the most effective and efficient use of scarce government-survey space, and develop survey measures that are both valid and reliable and that best complement and supplement existing regularly conducted surveys. They also must stay abreast of developments in the survey research field, including threats to traditional survey modes and

---

[21] This webpage describing their work on "word use, personality and well-being" can be found at http://www.cis.upenn.edu/~ungar/CVs/WWBP.html [October 2013].

[22] See http://www.knowledgenetworks.com [October 2013].

developments in alternative survey modes such as online surveys. Finally, for government data collection to stay relevant and feasible, statistical agencies will need to apportion some of their resources to following and understanding (and hopefully applying their own considerable expertise to) emerging methods of research designed to explore the use of both digital and digitized big data and other computational science methods for measuring people's behavior, attitudes, and states of well-being.

References

Abraham, K.G., S. Presser, and S. Helms. (2009). How social processes distort measurement: The impact of survey nonresponse on estimates of volunteer work in the United States. *American Journal of Sociology* 114(4):1129-1165.

Aguiar, M., and E. Hurst. (2007). Measuring trends in leisure: The allocation of time over five decades. *The Quarterly Journal of Economics* 122(3):969-1006.

Akerlof, G., and R. Kranton. (2002). Identity and schooling: Some lessons for the economics of education. *Journal of Economic Literature* 40(4):1167-1201.

Akerlof, G., and R. Kranton. (2010). *Identity Economics: How Our Identities Shape Our Work, Wages, and Well-Being*. Princeton: Princeton University Press.

Amato, P.R. (1999). Children of divorced parents as young adults. Pp. 147-163 in *Coping with Divorce. Single Parenting, and Marriage: A Risk and Resiliency Perspective*, E.M. Hetherington, ed. Mahwah, NJ: Lawrence Erlbaum.

Andrews, F., and S. Withey. (1976). *Social Indicators of Well-Being*. New York: Plenum Press.

Antenucci, D., M. Cafarella, M. Levenstein, and M. Shapiro. (2012). *Creating Measures of Labor Market Flows Using Social Media*. Presentation to the National Bureau of Economic Research, Cambridge, MA, July 16.

Baars, B.J. (2010). Spontaneous repetitive thoughts can be adaptive: Postscript on McKay and Vane. *Psychological Bulletin* 136(2):208-210.

Bertrand, M., and S. Mullainathan. (2001). Do people mean what they say? Implications for subjective survey data. *American Economic Review* 91(2):67-72.

Blattman, C., N. Fiala, and S. Martinez. (2013). *The Economic and Social Returns to Cash Transfers: Evidence from a Ugandan Aid Program*. Working paper. Available: http://cega.berkeley.edu/assets/cega_events/53/WGAPE_Sp2013_Blattman.pdf [June 2013].

Boehm, J.K., and L.D. Kubzansky. (2012). The heart's content: The association between positive psychological well-being and cardiovascular health. *Psychological Bulletin* 138(4):655-691.

Borton, J., L. Markowitz, and J. Dieterich. (2005). Effects of suppressing negative self-referent thoughts on mood and self-esteem. *Journal of Social and Clinical Psychology* 24(2):172-190.

Bradburn, N. (1968). Dimensions of marriage happiness. *American Journal of Sociology* 73(6):715-731.
Bradburn, N. (1969). *The Structure of Psychological Well-Being*. Chicago: Aldine.
Bradburn, N., and S.R. Orden. (1969). Working wives and marriage happiness. *American Journal of Sociology* 74(4):392-407.
Brickman, P., and D.T. Campbell. (1971). Hedonic relativism and planning the good society. Pp. 287-305 in *Adaptation-level Theory: A Symposium*, M.H. Appley, ed. New York: Academic Press.
Brickman, P., D. Coates, and R. Janoff-Bulman. (1978). Lottery winners and accident victims: Is happiness relative? *Journal of Personality and Social Psychology* 36(8):917-927.
Broderick, J.E., J.E. Schwartz, S. Schneider, and A.A. Stone. (2009). Can end-of-day reports replace momentary assessment of pain and fatigue? *The Journal of Pain* 10(3):274-281.
Brodeur, A. (2012). *Smoking, Income, and Subjective Well-Being: Evidence from Smoking Bans* (PSE working paper, no. 2012-03). Paris: Paris School of Economics.
Brodeur, A., and M. Connolly. (2012). *Do Higher Childcare Subsidies Improve Parental Well-Being? Evidence from Québec's Family Policies* (Cahiers de recherche 1223). Montreal, Canada: Centre Interuniversitaire sur le Risque, les Politiques Économiques et l'Emploi. Available: http://ideas.repec.org/p/lvl/lacicr/1223.html [October 2013].
Brülde, B. (2010). Happiness, morality, and politics. *Journal of Happiness Studies* 11(5): 567-583.
Burchell, B., D. Day, M. Hudson, D. Ladipo, R. Mankelow, J. Nolan, H. Reed, I.C. Wichert, and F. Wilkinson. (1999). *Job Insecurity and Work Intensification*. York, UK: Joseph Rowntree Foundation.
Cantril, H. (1965). *The Pattern of Human Concerns*. New Brunswick, NJ: Rutgers University Press.
Capps, C., and T. Wright. (2013). Toward a vision: Official statistics and big data. *Amstat News* The Membership Magazine of the American Statistical Association. Available: http://magazine.amstat.org/blog/2013/08/01/official-statistics [November 2013].
Carstensen, L.L. (2006). The influence of a sense of time on human development. *Science* 312(5782):1913-1915.
Charles, S.T., M. Mather, and L.L. Carstensen. (2003). Aging and emotional memory: The forgettable nature of negative images for older adults. *Journal of Experimental Psychology: General* 132(2):310-324.
Christodoulou, C., S. Schneider, and A.A. Stone. (2013). Validation of a brief yesterday measure of hedonic well-being and daily activities: Comparison with the Day Reconstruction Method. *Social Indicators Research*. Available: http://link.springer.com/content/pdf/10.1007%2Fs11205-013-0240-z.pdf [November 2012].
Clark, A.E., and C. Senik. (in press). The great happiness moderator. In *Essays on Inequality and Well-Being*, A.E. Clark and C. Senik, eds. Oxford, UK: Oxford University Press.
Clark, A.E., E. Diener, Y. Georgellis, and R.E. Lucas. (2008). Lags and leads in life satisfaction: A test of the baseline hypothesis. *The Economic Journal* 118(529):F222-F243.
Cohen, S., W.J. Doyle, R.B. Turner, C.M. Alper, and D.P. Skoner. (2003). Emotional style and susceptibility to the common cold. *Psychosomatic Medicine* 65(4):652-657.
Daly, M.C., and C. Gardiner. (in press). Disability and subjective well-being. In *Unexpected Lifecycle Events and Economic Well-Being*, K. Couch, M.C. Daly, and J. Zissimopoulus, eds. Redwood City, CA: Stanford University Press.
Davidson, R.J. (2004). Well-being and affective style: Neural substrates and biobehavioural correlates. *Philosophical Transactions of the Royal Society of London Series B: Biological Sciences* 359(1449):1395-1411.
Deaton, A. (2012). The financial crisis and the well-being of Americans. *Oxford Economic Papers* 64(1):1-26.

Deaton, A., and A.A. Stone. (2013a). *Evaluative and Hedonic Well-Being Among Those With and Without Children at Home*. Unpublished manuscript, Department of Economics, Princeton University.

Deaton, A., and A.A. Stone. (2013b). Two happiness puzzles. *American Economic Review* 103(3):591-597.

Di Tella, R., R.J. MacCulloch, and A.J. Oswald. (2001). Preferences over inflation and unemployment: Evidence from surveys of happiness. *American Economic Review* 91(1):335-341.

Diener, E. (2006). Guidelines for national indicators of subjective well-being and ill-being. *Applied Research in Quality of Life* 1(2):151-157.

Diener, E. (2011). *The Validity of Life Satisfaction Measures*. Unpublished manuscript, Department of Psychology, University of Illinois at Urbana-Champaign.

Diener, E., and M. Chan. (2010). *Happier People Live Longer: Subjective Well-Being Contributes to Health and Longevity*. Unpublished manuscript, Department of Psychology, University of Illinois at Urbana-Champaign.

Diener, E., and M.E.P. Seligman. (2004). Beyond money: Toward an economy of well-being. *Psychological Science in the Public Interest*, 5(1). Available: http://internal.psychology.illinois.edu/~ediener/Documents/Diener-Seligman_2004.pdf [November 2013].

Diener, E., and L. Tay. (2013). Review of the Day Reconstruction Method (DRM). *Social Indicators Research*. Available: http://download.springer.com/static/pdf/569/art%253A10.1007%252Fs11205-013-0279-x.pdf?auth66=1380465190_fb2a9ce814c8464852cefd83ac729868&ext=.pdf [July 2013].

Diener, E., R. Emmons, R.J. Larsen, and S. Griffin. (1985). The satisfaction with life scale. *Journal of Personality Assessment* 49(1):71-75.

Diener, E., H. Smith, and F. Fujita. (1995). The personality structure of affect. *Journal of Personality and Social Psychology* 69(1):130-141.

Diener, E., E.M. Suh, R.E. Lucas, and H.L. Smith. (1999). Subjective well-being: Three decades of progress. *Psychological Bulletin* 125(2):276-302.

Diener, E., D. Kahneman, W. Tov, R. Arora, and J. Harter. (2009). Income's differential influence on judgments of life versus affective well-being. Pp. 233-246 in *Assessing Well-Being*, E. Diener, ed. Oxford, UK: Springer.

Diener, E., W. Ng, J. Harter, and R. Arora. (2010). Wealth and happiness across the world: Material prosperity predicts life evaluation, whereas psychosocial prosperity predicts positive feeling. *Journal of Personality and Social Psychology* 99(1):52-61.

Diener, E., R. Inglehart, and L. Tay. (2013). Theory and validity of life satisfaction scales. *Social Indicators Research* 112(3):497-527.

Dolan, P. (2008). Developing methods that really do value the "Q" in the QALY. *Health Economics, Policy, and Law* 3(1):69-78.

Dolan, P. (2011). Thinking about it: Thoughts about health and valuing QALYs. *Health Economics* 20(12):1407-1416.

Dolan, P. (2012). *NAS SWB Position Paper*. Background paper for NAS Panel on Well-Being Metrics. Washington, DC.

Dolan, P., and D. Kahneman. (2008). Interpretations of utility and their implications for the valuation of health. *The Economic Journal* 118(525):215-234.

Dolan, P., and G. Kavetsos. (2012). *Happy Talk: Mode of Administration Effects on Subjective Well-Being* (CEP discussion paper, no. 1159). London, UK: Centre for Economic Performance, London School of Economics and Political Science. Available: http://eprints.lse.ac.uk/45273 [November 2012].

Dolan, P., and R. Metcalfe. (2008). *Comparing Willingness-to-Pay and Subjective Well-Being in the Context of Non-Market Goods* (CEP discussion paper, no. 890). London, UK: Centre for Economic Performance, London School of Economics and Political Science.

Dolan, P., and R. Metcalfe. (2010). Oops . . . I did it again: Repeated focusing effects in reports of happiness. *Journal of Economic Psychology* 31(4):732-737.

Dolan, P., and R. Metcalfe. (2011). *Comparing Measures of Subjective Well-Being and Views About the Role They Should Play in Policy*. London, UK: Office for National Statistics.

Dolan, P., G. Kavetsos, and A. Tsuchiya. (2013). Sick but satisfied: The impact of life and health satisfaction on choice between health scenarios. *Journal of Health Economics* 32(4)708-714.

Durlauf, S.N. (2002). On the empirics of social capital. *Economic Journal: Royal Economic Society* 112(483):459-479.

Easterlin, R.A. (2001). Income and happiness: Towards a unified theory. *The Economic Journal* 111(473):465-484.

Easterlin, R.A. (2004). Explaining happiness. *Proceedings of the National Academy of Sciences of the United States of America* 100(19):1176-1183.

Easterlin, R.A. (2005). Diminishing marginal utility of income? A caveat emptor. *Social Indicators Research* 70(3):243-255.

Eid, M., and E. Diener. (2004). Global judgments of subjective well-being: Situational variability and long-term stability. *Social Indicators Research* 65(3):245-277.

Ersner-Hershfield, H., J.A. Mikels, S.J. Sullivan, and L.L. Carstensen. (2008). Poignancy: Mixed emotional experience in the face of meaningful endings. *Journal of Personality and Social Psychology* 94(1):158-167.

Feldman, F. (2004). *Pleasure and the Good Life*. Oxford, UK: Clarendon Press.

Fernandes, M., M. Ross, M. Wiegand, and E. Schryer. (2008). Are the memories of older adults positively biased? *Psychology and Aging* 23(2):297-306.

Ferreira, S., and M. Moro. (2009). *On the Use of Subjective Well-Being Data for Environmental Valuation* (Stirling economics discussion papers 2009-24). Stirling, UK: Division of Economics, University of Stirling.

Ferrie, J., M. Shipley, S. Stansfeld, and M. Marmot. (2002). Effects of chronic job insecurity and change in job security on self-reported health, minor psychiatric morbidity, physiological measures, and health related behaviours in British civil servants: The Whitehall II study. *Journal of Epidemiology and Community Health* 56(6):450-454.

Fredrickson, B.L., K.M. Grewer, K.A. Coffey, S.B. Algoe, A.M. Firestine, J.M.G. Arevalo, J. Ma, and S.W. Cole. (2013). A functional genomic perspective on human well-being. *Proceedings of the National Academy of Sciences of the United States of America*. Available: http://www.pnas.org/content/early/2013/07/25/1305419110.full.pdf+html [August 2013].

Fujiwara, D., and R. Campbell. (2011). *Valuation Techniques for Social Cost-Benefit Analysis: Stated Preference, Revealed Preference and Subjective Well-Being Approaches. A discussion of the current issues*. London, UK: Her Majesty's Treasury; Department of Works and Pensions.

Fulmer, C.A., M.J. Gelfand, A.W. Kruglanski, C. Kim-Prieto, E. Diener, A. Pierro, A., and E.T. Higgins. (2010). Feeling "all right" in societal contexts: Person-culture trait match and its impact on self-esteem and subjective well-being. *Psychological Science* 21(11):1563-1569.

Gandelman, N., and R. Hernández-Murillo. (2009). The impact of inflation and unemployment on subjective personal and country evaluations. *Federal Reserve Bank of St. Louis Review* 91(3):107-126.

Garau, M., K.K. Shah, A.R. Mason, Q. Wang, A. Towse, and M.F. Drummond. (2011). Using QALYs in cancer: A review of the methodological limitations. *Pharmacoeconomics* 29(8):673-685.

Gere, J., and U. Schimmack. (2011). A multi-occasion multi-rater model of affective dispositions and affective well-being. *Journal of Happiness Studies* 12(6):931-945.

REFERENCES

Graham, C. (2008). Happiness and health: Lessons—and questions—for public policy. *Health Affairs* 27(1):72-87.
Graham, C. (2011). *The Pursuit of Happiness: An Economy of Well-Being*. Washington, DC: The Brookings Institution Press.
Graham, C., and E. Lora, eds. (2009). *Paradox and Perception: Measuring Quality of Life in Latin America*. Washington, DC: The Brookings Institution Press.
Graham, C., and J. Markowitz. (2011). Aspirations and happiness of potential Latin American immigrants. *Journal of Social Research and Policy* 2(2):9-25.
Graham, C., and S. Pettinato. (2002). *Happiness and Hardship: Opportunity and Insecurity in New Market Economies*. Washington, DC: The Brookings Institution Press.
Graham, C., L. Higuera, and E. Lora. (2011). Which health conditions cause the most unhappiness? *Health Economics* 20(12):1431-1447.
Greenfield, E.A., and N.F. Marks. (2004). Formal volunteering as a protective factor for older adults' psychological well-being. *Journals of Gerontology Series B: Psychological Sciences and Social Sciences* 59(5):S258-S264.
Harmon-Jones, E. (2004). On the relationship of anterior brain activity and anger: Examining the role of attitude toward anger. *Cognition and Emotion* 18(3):337-361.
Hay, J.L., K.D. Mccaul, and R.E. Magnan. (2006). Does worry about breast cancer predict screening behaviors? A meta-analysis of the prospective evidence. *Preventive Medicine* 42(6):401-408.
Heckman, J.J. (2000). Causal parameters and policy analysis in economics: A twentieth century retrospective. *Quarterly Journal of Economics* 115(1):45-97. Available: http://www.jstor.org/stable/2586935 [March 2013].
Heckman, J.J., J. Stixrud, and S. Urzua. (2006). The effects of cognitive and noncognitive abilities on labor market outcomes and social behavior. *Journal of Labor Economics* 24(3):411-482.
Helliwell, J.F. (2002). How's life? Combining individual and national variables to explain subjective well-being. *Economic Modelling* 20(2003):331-360. Available: http://faculty.arts.ubc.ca/jhelliwell/papers/Helliwell-EM2003.pdf [October 2012].
Helliwell, J.F., and R.D. Putnam. (2004). The social context of well-being. *Philosophical Transactions of the Royal Society of London Series B: Biological Sciences* 359(1449):1435-1446.
Helliwell, J.F., R. Layard, and J. Sachs, eds. (2012). *World Happiness Report*. New York: Earth Institute, Columbia University. Available: http://issuu.com/earthinstitute/docs/world-happiness-report?e=4098028/2014244 [February 2013].
Hershfield, H.E., S. Scheibe, T. Sims, and L.L. Carstensen. (2013). When feeling bad can be good: Mixed emotions benefit physical health across adulthood. *Social Psychological and Personality Science* 4(1):54-61.
Huppert, F.A. (2009). Psychological well-being: Evidence regarding its causes and consequences. *Applied Psychology: Health and Well-being* 1(2):137-164.
Huppert, F.A., N. Baylis, and B. Keverne. (2004). Introduction: Why do we need a science of well-being? *Philosophical Transactions of the Royal Society of London Series B: Biological Sciences* 359(1449):1331-1332.
Hyde, M., R.D. Wiggins, P. Higgs, and D.B. Blane. (2003). A measure of quality of life in early old age: The theory, development, and properties of a needs satisfaction model (CASP-19). *Aging and Mental Health* 7(3):86-94.
Kahneman, D. (1999). Objective happiness. Pp. 3-25 in *Well-Being: Foundations of Hedonic Psychology*, D. Kahneman, E. Diener, and N. Schwarz, eds. New York: Russell Sage Foundation Press.
Kahneman, D., and A. Deaton. (2010). High income improves evaluation of life but not emotional well-being. *Proceedings of the National Academy of Sciences of the United States of America* 107(38):16489-16493.

Kahneman, D., and A.B. Krueger. (2006). Developments in the measurement of subjective well-being. *Journal of Economic Perspectives* 20(1):3-24.

Kahneman, D., and D. Miller. (1986). Norm theory: Comparing reality to its alternative. *Psychology Review* 93(2):136-153.

Kahneman, D., A.B. Krueger, D. Schkade, N. Schwarz, and A.A. Stone. (2004). A survey method for characterizing daily life experience: The Day Reconstruction Method (DRM). *Science* 306(5702):1776-1780.

Kahneman, D., A.B. Krueger, D. Schkade, N. Schwarz, and A.A. Stone. (2006). Would you be happier if you were richer? A focusing illusion. *Science* 312(5782):1908-1910.

Kapteyn, A., J.P. Smith, and A. Van Soest. (2010). Life satisfaction. Pp. 70-104 in *International Differences in Well-Being*, E. Diener, J.F. Helliwell, and D. Kahneman, eds. Oxford, UK: Oxford University Press.

Kapteyn, A., J. Lee, C. Tassot, H. Vonkova, and G. Zamarro. (2013). *Dimensions of Subjective Well-Being* (CESR working paper series, paper no. 2013-005). Playa Vista, CA: Dornsife Center for Economic and Social Research. Available: https://cesr.usc.edu/documents/WP_2013_005.pdf [July 2013].

Kennedy, Q., M. Mather, and L.L. Carstensen. (2004). The role of motivation in the age-related positivity effect in autobiographical memory. *Psychological Science* 15(3):208-214.

Kessler, E.M., and U.M. Staudinger. (2009). Affective experience in adulthood and old age: The role of affective arousal and perceived affect regulation. *Psychology and Aging* 24(2):349-362.

Keyes, C.L.M. (2002). The mental health continuum: From languishing to flourishing in life. *Journal of Health and Social Behavior* 43(2):207-222.

Killingsworth, M.A., and D.T. Gilbert. (2010). A wandering mind is an unhappy mind. *Science* 330(6006):932.

Kim, H.S., J.Y. Park, and B.B. Jin. (2008). Dimensions of online community attributes: Examination of online communities hosted by companies in Korea. *International Journal of Retail and Distribution Management* 36(10):812-830.

Kominski, R., and K. Short. (1996). *Developing Extended Measures of Well-Being: Minimum Income and Subjective Income Assessments* (SIPP working paper no. 228). Washington, DC: U.S. Bureau of the Census.

Kristensen, N., and Johansson, E. (2008). New evidence on cross-country differences in job satisfaction using anchoring vignettes. *Labour Economics* 15(1):96-117.

Krueger, A.B., and A. Mueller. (2008). *The Lot of the Unemployed: A Time Use Perspective* (IZA discussion paper no. 3490). Available: http://ftp.iza.org/dp3490.pdf [October 2013].

Krueger, A.B., and A. Mueller. (2011). *Job Search, Emotional Well-Being, and Job Finding in a Period of Mass Unemployment: Evidence from High Frequency Longitudinal Data*. Washington, DC: Brookings Papers on Economic Activity.

Krueger, A.B., and A. Mueller. (2012). Time use, emotional well-being, and unemployment: Evidence from longitudinal data. *American Economic Review* 102(3):594-599.

Krueger, A.B., and D. Schkade. (2008). The reliability of subjective well-being measures. *Journal of Public Economics* 92(8-9):1833-1845.

Krueger, A.B., and A.A. Stone. (2008). Assessment of pain: A community-based diary survey in the USA. *Lancet* 371(9623):1519-1525.

Krueger, A.B., D. Kahneman, D. Schkade, N. Schwarz, and A.A. Stone. (2009). National time accounting: The currency of life. Pp. 9-86 in *Measuring the Subjective Well-Being of Nations: National Accounts of Time Use and Well-Being*, A.B. Krueger, ed. Chicago: Chicago University Press.

Larson, R.J., and E. Diener. (1992). Promises and problems with the circumplex view of emotion. Pp. 25-59 in *Review of Personality and Social Psychology: Emotion*, M.S. Clark, ed. Newbury Park, CA: Sage.

Layard, R. (2006). Happiness and public policy: A challenge to the profession. *Economic Journal* 116(510):C24-C33.

Leyden, K.M., A. Goldberg, and P. Michelbach. (2011). Understanding the pursuit of happiness in ten major cities. *Urban Affairs Review* 47(6):861-888.

Löckenhoff, C.E., and L.L. Carstensen. (2007). Aging, emotion, and health-related decision strategies: Motivational manipulations can reduce age differences. *Psychology and Aging* 22(1):134-146.

Löckenhoff, C.E., and L.L. Carstensen. (2008). Decision strategies in healthcare choices for self and others: Older adults make adjustments for the age of the decision target, younger adults do not. *Journals of Gerontology Series B: Psychological Sciences and Social Sciences* 63(2):106-109.

Loewenstein, G., and P.A. Ubel. (2008). Hedonic adaptation and the role of decision and experience utility in public policy. *Journal of Public Economics* 92(8-9):1795-1810.

Loomes, G., and L. McKenzie. (1989). The use of QALYs in health care decision making. *Social Science and Medicine* 28(4):299-308.

Lucas, R.E., E. Diener, and E. Suh. (1996). Discriminant validity of well-being measures. *Journal of Personality and Social Psychology* 71(3):616-628.

Lucas, R.E., A.E. Clark, Y. Georgellis, and E. Diener. (2003). Reexamining adaptation and the set point model of happiness: Reactions to changes in marital status. *Journal of Personality and Social Psychology* 84(3):527-539.

Lucas, R.E., A.E. Clark, Y. Georgellis, and E. Diener. (2004). Unemployment alters the set point for life satisfaction. *Psychological Science* 15(1):8-13.

Luechinger, S. (2009). Valuing air quality using the life satisfaction approach. *Economic Journal* 119(536):482-515.

Luhmann, M., U. Schimmack, and M. Eid. (2011). Stability and variability in the relationship between subjective well-being and income. *Journal of Research in Personality* 45(2):186-197.

MacKerron, G., and S. Mourato. (2013). Happiness is greater in natural environments. In *Global Environmental Change*. Available: http://www.sciencedirect.com/science/article/pii/S0959378013000575 [July 2013].

Mather, M., and M.K. Johnson. (2000). Choice-supportive source monitoring: Do our decisions seem better to us as we age? *Psychology and Aging* 15(4):596-606.

Mather, M., M. Knight, and M. McCaffrey. (2005). The allure of the alignable: Younger and older adults' false memories of choice features. *Journal of Experimental Psychology: General* 134(1):38-51.

May, J., J. Andrade, K. Willoughby, and C. Brown. (2011). *An Attentional Control Task Reduces Intrusive Thoughts About Smoking*. Oxford, UK: Oxford University Press.

Meier, S., and A. Stutzer. (2006). *Is Volunteering Rewarding in Itself?* Center for Behavioral Economics and Decision-Making, Federal Reserve Bank of Boston.

Metcalfe, R., N. Powdthavee, and P. Dolan. (2011). Destruction and distress: Using a quasi-experiment to show the effects of the September 11 attacks on mental well-being in the United Kingdom. *Economic Journal* 121(550):F81-F103.

Meyer, B.D., and J.X. Sullivan. (2009). *Economic Well-Being and Time Use*. Working paper, June 22. Available: http://www.sole-jole.org/12460.pdf [October 2013].

Mikels, J.A., G.R. Larkin, P.A. Reuter-Lorenz, and L.L. Carstensen. (2005). Divergent trajectories in the aging mind: Changes in working memory for affective versus visual information with age. *Psychology and Aging* 20(4):542-553.

National Research Council. (2013). *Principles and Practices for a Federal Statistical Agency, Fifth Edition.* C.F. Citro and M.L. Straf, eds. Committee on National Statistics. Division of Behavioral and Social Sciences and Education. Washington, DC: The National Academies Press.

OECD. (2013). *OECD Guidelines on Measuring Subjective Well-being.* Paris: OECD. Available: http://dx.doi.org/10.1787/9789264191655-en [October 2013].

Office for National Statistics. (2011). *Initial Investigation into Subjective Well-being from the Opinions Survey.* Newport, UK: Office for National Statistics. Available: http://www.ons.gov.uk/ons/dcp171776_244488.pdf [October 2013].

Office for National Statistics. (2013). *Personal Well-Being in the UK, 2012/13* (Statistical Bulletin). Newport, UK: Office for National Statistics. Available: http://www.ons.gov.uk/ons/rel/wellbeing/measuring-national-well-being/personal-well-being-in-the-uk--2012-13/sb---personal-well-being-in-the-uk--2012-13.html [September 2013].

Oishi, S., U. Schimmack, and S.J. Colcombe. (2003). The contextual and systematic nature of life satisfaction judgments. *Journal of Experimental Social Psychology* 39(2003):232-247.

Oswald, A.J., and S. Wu. (2009). *Well-Being Across America: Evidence from a Random Sample of One Million U.S. Citizens.* Unpublished manuscript, University of Warwick, UK. Presented at the IZA Prize Conference, October 22, Washington, DC. Available: http://www.iza.org/conference_files/prizeconf2009/oswald_a262.pdf [November 2012].

Passel, J.S., and D.-V, Cohn. (2008). *U.S. Population Projections: 2005-2050.* Pew Research Center, February 11. Available: http://pewhispanic.org/files/reports/85.pdf [September 2013].

Quericia, D., J. Ellis, L. Capra, and J. Crowcroft. (2012). Tracking "gross community happiness" from tweets. Pp. 965-968 in *CSW '12: Proceedings of the ACM 2012 Conference on Computer Supported Cooperative Work.* New York: ACM. Available: http://dl.acm.org/citation.cfm?id=2145347 [October 2013].

Redelmeier, D.A., and D. Kahneman. (1996). Patients' memories of painful medical treatments: Real-time and retrospective evaluations of two minimally invasive procedures. *Pain* 66(1):3-8.

Reed, A. E., and L.L. Carstensen. (2012). The theory behind the age-related positivity effect. *Frontiers in Psychology* 3(339):1-9.

Riis, J., G. Lowenstein, J. Baron, and C. Jepson. (2005). Ignorance of hedonic adaptation: A study using ecological momentary assessment. *Journal of Experimental Psychology: General* 134(1):3-9.

Robinson, M.D., and G.L. Clore. (2002). Belief and feeling: Evidence for an accessibility model of emotional self-report. *Psychological Bulletin* 128(6):934-960.

Ryff, C., and C. Keyes. (1995). The structure of psychological well-being revisited. *Journal of Personality and Social Psychology* 69(4):719-727.

Sampson, R.J., and C. Graif. (2009). Neighborhood social capital as differential social organization: Resident and leadership dimensions. *American Behavioral Scientist* 52(11):1579-1605.

Schimmack, U. (2008). The structure of subjective well-being. Pp. 97-123 in *The Science of Subjective Well-Being,* M. Eid and R. Larsen, eds. New York: Guilford Press.

Schimmack, U., and S. Oishi. (2005). The influence of chronically and temporarily accessible information on life satisfaction judgments. *Journal of Personality and Social Psychology* 89(3):395-406.

Schlagman, S., J. Schulz, and L. Kvavilashvili. (2006). A content analysis of involuntary autobiographical memories: Examining the positivity effect in old age. *Memory* 14(2):161-175.

Schneider, S., A.A. Stone, J.E. Schwartz, and J.E. Broderick. (2011). Peak and end effects in patients' daily recall of pain and fatigue: A within-subjects analysis. *The Journal of Pain* 12(2):228-235.

Schuller, T., M. Wadsworth, J. Bynner, and H. Goldstein. (2012). *The Measurement of Well-being: The Contribution of Longitudinal Studies*. Report prepared for the Office for National Statistics. London. UK: Longview. Available: http://www.longviewuk.com/pages/documents/Longviewwellbeingreport.pdf [October 2013].

Schwarz, N. (1987). *Stimmung als Information: Untersuchungen zum Einfluß von Stimmungen auf die Bewertung des eigenen Lebens*. Heidelberg: Springer Verlag.

Schwarz, N., and H. Schuman. (1997). Political knowledge, attribution, and inferred interest in politics: The operation of buffer items. *International Journal of Public Opinion Research* 9(2):191-195.

Schwarz, N., and F. Strack. (1999). Reports of subjective well-being: Judgmental processes and their methodological implications. Pp. 61-84 in *Well-Being: The Foundations of Hedonic Psychology*, D. Kahneman, E. Diener, and N. Schwarz, eds. New York: Russell-Sage. Available: http://sitemaker.umich.edu/norbert.schwarz/files/99_wb_schw_strack_reports_of_wb.pdf [January 2013].

Schwarz, N., D. Kahneman, and J. Xu. (2009). Global and episodic reports of hedonic experience. Pp. 157-174 in *Using Calendar and Diary Methods in Life Events Research*, R. Belli, D. Alwin, and F. Stafford, eds. Newbury Park, CA: Sage.

Seaford, C. (2011). Policy: Time to legislate for the good life. *Nature* 477(7366):532-533.

Seligman, M., and M. Csikszentmihalyi. (2000). Positive psychology: An introduction. *American Psychologist* 55(1):5-14.

Sen, A.K. (1985). *Commodities and Capabilities*. Amsterdam, Netherlands: Elsevier.

Shiffman, S., and A.A. Stone. (1998). Ecological Momentary Assessment: A new tool for behavior medicine research. Pp. 117-131 in *Technology and Methods in Behavioral Medicine*, D.S. Krants, ed. Mahwah, NJ: Lawrence Erlbaum Associates.

Smallwood, J., and J.W. Schooler. (2006). The restless mind. *Psychological Bulletin* 132(6):946-958.

Smith, D.M. (2012). *Ecological Momentary Assessment and the Day Reconstruction Method*. Paper commissioned by the Panel on Subjective Well-Being in a Policy-Relevant Framework. Unpublished manuscript, Committee on National Statistics, National Research Council, Washington, DC.

Smith, N.K., J.T. Larsen, R.L. Chartrand, J.T. Cacioppo, H.A. Katafiaz, and K.E. Moran. (2006). Being bad isn't always good: Affective context moderates the attention bias toward negative information. *Journal of Personality and Social Psychology* 90(4):210-220.

Smith, T.W. (2005). *Troubles in America: A Study of Negative Life Events Across Time and Sub-Groups* (Russell Sage Foundation Working Paper Series). Available: http://www.russellsage.org/research/reports/troubles-in-america [October 2013].

Springer, K.W., R.M. Hauser, and J. Freese. (2006). Reply: Bad news indeed for Ryff's six-factor model of well-being. *Social Science Research* 35(4):1120-1131.

Steg, L., and Gifford, R. (2005). Sustainable transportation and quality of life. *Journal of Transport Geography* 13(1):59-69.

Steptoe, A., and J. Wardle. (2011). Positive affect measured using ecological momentary assessment and survival in older men and women. *Proceedings of the National Academy of Sciences of the United States of America* 108(45):18244-18248.

Steptoe, A., J. Wardle, and M. Marmot. (2005). Positive affect and health-related neuroendocrine, cardiovascular, and inflammatory processes. *Proceedings of the National Academy of Sciences of the United States of America* 102(18):6508-6512.

Stevenson, B., and J. Wolfers. (2008). Economic growth and subjective well-being: Reassessing the Easterlin paradox. *Brookings Papers on Economic Activity* 39(1):1-102.

Stevenson, B., and J. Wolfers. (2013). Subjective well-being and income: Is there any evidence of satiation? *American Economic Review* 103(3):598-604.

Stiglitz, J.E., A. Sen, and J.-P. Fitoussi. (2009). *Report by the Commission on the Measurement of Economic Performance and Social Progress.* Available: http://www.stiglitz-sen-fitoussi.fr/documents/rapport_anglais.pdf [August 2012].

Stone, A.A. (2011). A rationale for including a brief assessment of hedonic well-being in large-scale surveys. *Forum for Health Economics and Policy* 14(3), article 7.

Stone, A.A., and S. Shiffman. (1994). Ecological Momentary Assessment (EMA) in behavioral medicine. *Annals of Behavioral Medicine* 16(3)199-202.

Stone, A.A., J.E. Broderick, A.T. Kaell, P.A. DelesPaul, and L.E. Porter. (2000). Does the peak-end phenomenon observed in laboratory pain studies apply to real-world pain in rheumatoid arthritis? *Journal of Pain* 1(3):212-217.

Stone, A.A., J.E. Schwartz, N. Schwarz, D. Schkade, A. Krueger, and D. Kahneman. (2006). A population approach to the study of emotion: Diurnal rhythms of a working day examined with the Day Reconstruction Method (DRM). *Emotion* 6(5):139-149.

Stone, A.A., J.E., Schwartz, J.E., Broderick, and A. Deaton. (2010). A snapshot of the age distribution of psychological well-being in the United States. *Proceedings of the National Academy of Sciences of the United States of America* 107(22):9985-9990.

Stone, A.A., S. Schneider, and J.K. Harter. (2012). Day-of-week mood patterns in the United States: On the existence of "blue Monday," "thank God it's Friday," and weekend effects. *Journal of Positive Psychology* 7(12):306-314.

Stutzer, A., and B.S. Frey. (2004). Reported subjective well-being: A challenge for economic theory and economic policy. *Schmollers Jahrbuch: Zeitschrift für Wirtschafts und Sozialwissenschaften* 124(2):191-231.

Su, R., L. Tay, and E. Diener. (2013). Scales to assess eudaimonic and subjective well-being. Paper in preparation.

Taylor, M.P. (2006). Tell me why I don't like Mondays: Investigating day-of-the-week effects on job satisfaction and psychological well-being. *Journal of the Royal Statistical Society Series A* 169(1):127-142.

Tellegen, A., D. Watson, and L.A. Clark. (1994). Modeling dimensions of mood. In L.A. Feldman (chair), *Mood: Consensus and Controversy.* Symposium presented at the 102nd Annual Convention of the American Psychological Association, Los Angeles, CA.

Thoits, P. (1983). Multiple identities and psychological well-being. *American Sociological Review* 48(2):174-182.

Tsai, J.L. (2007). Ideal affect: Cultural causes and behavioral consequences. *Perspectives on Psychological Science* 2:242-259.

Tsai, J.L., B. Knutson, and H.H. Fung. (2006). Cultural variation in affect valuation. *Journal of Personality and Social Psychology* 90(2):288-307.

Ubel, P.A., Y. Peeters, and D. Smith. (2010). Abandoning the language of "response shift": A plea for conceptual clarity in distinguishing scale recalibration from true changes in quality of life. *Quality of Life Research* 19(4):465-471.

Van Soeste, A., L. Delaney, C. Harmon, A. Kapteyn, and J.P. Smith. (2011). Validating the use of anchoring vignettes for the correction of response scale differences in subjective questions. *Journal of the Royal Statistical Society: Series A (Statistics in Society)* 174(3):575-595.

Vanhoutte, B., J. Nazroo, and T. Chandola. (2012). *Measuring Well-Being in Later Life.* Paper presented at the SHARE Users Conference, July 28-29, Ca' Foscari University, Venice, Italy.

Watkins, E. (2008). Constructive and unconstructive repetitive thought. *Psychological Bulletin* 134(2):163-206.

Watson, D., and L.A. Clark. (1999). *Manual for the Positive and Negative Affect Schedule—Expanded Form.* Iowa City: University of Iowa.

Watson, D., L.A. Clark, and A. Tellegen. (1988). Development and validation of brief measures of positive and negative affect: The PANAS scales. *Journal of Personality and Social Psychology* 54(6):1063-1070.

Wernick, J. (2008). *Building Happiness: Architecture to Make You Smile*. London, UK: Black Dog.

White, M.P., and P. Dolan. (2009). Accounting for the richness of daily activities. *Psychological Science* 20(8):1000-1008.

Zou, C., U. Schimmack, and J. Gere. (unpublished). *Towards an Integrated Theory of the Nature and Measurement of Well-being: A Multiple-Indicator-Multiple-Rater Model*. Manuscript submitted for publication, University of Toronto Mississauga.

Appendix A

Experienced Well-Being Questions and Modules from Existing Surveys

Included in this appendix are examples of subjective well-being (SWB) modules that have been used in various surveys. The first set is the UK Office for National Statistics SWB module used in the Integrated Household Survey.[1] The remaining three sets are experienced well-being (ExWB) questions compiled by Kapteyn and colleagues (2013, p. 10) from three sources:

1. The English Longitudinal Study of Ageing;
2. The Gallup-Healthways Well-Being Index; and
3. HWB-12, a set of 12 questions to assess hedonic well-being, which was developed by Jacqui Smith and Arthur Stone and included in the 2012 administration of the Health and Retirement Study.

These examples are meant to illustrate question wording and the scope of SWB modules; they are far from comprehensive. The Annexes in the OECD *Guidelines* (OECD, 2013) offer another set of examples of SWB measures and sample question modules that draw broadly from existing surveys.

---

[1] See http://www.ons.gov.uk/ons/rel/wellbeing/measuring-subjective-wellbeing-in-the-uk/first-annual-ons-experimental-subjective-well-being-results/first-ons-annual-experimental-subjective-well-being-results.html#tab-Background [October 2013].

UK OFFICE FOR NATIONAL STATISTICS SWB MODULE

Between April 2011 and March 2012, four subjective well-being questions were included in the constituent surveys of the Integrated Household Survey:

1. Overall, how satisfied are you with your life nowadays?
2. Overall, to what extent do you feel the things you do in your life are worthwhile?
3. Overall, how happy did you feel yesterday?
4. Overall, how anxious did you feel yesterday?

All were answered on a scale of *0* to *10* where *0* is "not at all" and *10* is "completely."

ExWB QUESTIONS FROM THE ENGLISH LONGITUDINAL SURVEY OF AGEING

What day of the week was it **yesterday?** *Tick one box.*
- ☐ Monday
- ☐ Tuesday
- ☐ Wednesday
- ☐ Thursday
- ☐ Friday
- ☐ Saturday
- ☐ Sunday

What time did you wake up **yesterday?** *For example, if you woke up at 4:00 AM, please enter 04 for the hour, 00 for the minutes, and circle AM.*

Hours___ Minutes___ AM or PM

What time did you go to sleep at the end of the day **yesterday?** *For example, if you went to sleep at 11:30 PM, please enter 11 for the hour, 30 for the minutes, and circle PM.*

Hours___ Minutes___ AM or PM

APPENDIX A

Yesterday, did you feel any **pain**?
None ☐
A little ☐
Some ☐
Quite a bit ☐
A lot ☐

Did you feel well-rested **yesterday morning** (that is, you slept well the night before)?
Yes ☐   No ☐

Was **yesterday** a normal day for you or did something unusual happen? *Tick one box.*
Yes, just a normal day ☐
No, my day included unusual bad (stressful) things ☐
No, my day included unusual good things ☐

Intro: Please think about the **things you did yesterday**. How did you spend your time and how did you feel?

Yesterday, did you **watch TV**? *Tick one box.*
Yes ☐
No ☐ (skip next 2 questions)

How much time did you spend **watching TV yesterday**? *For example, if you spent one and a half hours, enter 1 for the hours and 30 for the minutes.*
Hours___ Minutes___

How did you feel when you were **watching TV yesterday**? Rate each feeling on a scale from 0 (did not experience at all) to 6 (the feeling was extremely strong). *Tick one box on each line.*

I felt

	0	1	2	3	4	5	6
Happy	☐	☐	☐	☐	☐	☐	☐
Interested	☐	☐	☐	☐	☐	☐	☐
Frustrated	☐	☐	☐	☐	☐	☐	☐
Sad	☐	☐	☐	☐	☐	☐	☐

Yesterday, did you **work or volunteer**? *Tick one box.*
Yes ☐
No ☐ (skip next 2 questions)

How much time did you spend **working or volunteering yesterday**? *For example, if you spent nine and a half hours, enter 9 for the hours and 30 for the minutes.*

Hours___ Minutes___

How did you feel when you were **working or volunteering yesterday**? Rate each feeling on a scale from 0 (did not experience at all) to 6 (the feeling was extremely strong). *Tick one box on each line.*

I felt

	0	1	2	3	4	5	6
Happy	☐	☐	☐	☐	☐	☐	☐
Interested	☐	☐	☐	☐	☐	☐	☐
Frustrated	☐	☐	☐	☐	☐	☐	☐
Sad	☐	☐	☐	☐	☐	☐	☐

Yesterday, did you go for a **walk or exercise**? *Tick one box.*
Yes ☐
No ☐ (skip next 2 questions)

How much time did you spend **walking or exercising yesterday**? *For example, if you spent 30 minutes, enter 0 for the hours and 30 for the minutes.*
Hours___ Minutes___

How did you feel when you were **walking or exercising yesterday**? Rate each feeling on a scale from 0 (did not experience at all) to 6 (the feeling was extremely strong). *Tick one box on each line.*

I felt

	0	1	2	3	4	5	6
Happy	☐	☐	☐	☐	☐	☐	☐
Interested	☐	☐	☐	☐	☐	☐	☐
Frustrated	☐	☐	☐	☐	☐	☐	☐
Sad	☐	☐	☐	☐	☐	☐	☐

Yesterday, did you do any **health-related activities other than walking or exercise?** *For example, did you visit a doctor, take medications, or have a treatment? Tick one box.*
Yes ☐
No ☐ (skip next 2 questions)

How much time did you spend doing **health-related activities yesterday?**
Hours___ Minutes___

How did you feel when you were **doing health-related activities yesterday?** Rate each feeling on a scale from 0 (did not experience at all) to 6 (the feeling was extremely strong). *Tick one box on each line.*

I felt

	0	1	2	3	4	5	6
Happy	☐	☐	☐	☐	☐	☐	☐
Interested	☐	☐	☐	☐	☐	☐	☐
Frustrated	☐	☐	☐	☐	☐	☐	☐
Sad	☐	☐	☐	☐	☐	☐	☐

Yesterday, did you **travel or commute?** *For example, by car, train, bus, etc. Tick one box.*
Yes ☐
No ☐ (skip next 2 questions)

How much time did spend **traveling or commuting yesterday?**
Hours___ Minutes___

How did you feel when you were **traveling or commuting yesterday?** Rate each feeling on a scale from 0 (did not experience at all) to 6 (the feeling was extremely strong). *Tick one box on each line.*

I felt

	0	1	2	3	4	5	6
Happy	☐	☐	☐	☐	☐	☐	☐
Interested	☐	☐	☐	☐	☐	☐	☐
Frustrated	☐	☐	☐	☐	☐	☐	☐
Sad	☐	☐	☐	☐	☐	☐	☐

Yesterday, did you **spend time with friends or family?** *Tick one box.*
Yes ☐
No ☐ (skip next 2 questions)

How much time did you spend **with friends or family yesterday?**
Hours___ Minutes___

How did you feel when you were **with friends or family yesterday?** Rate each feeling on a scale from 0 (did not experience at all) to 6 (the feeling was extremely strong). *Tick one box on each line.*

I felt

	0	1	2	3	4	5	6
Happy	☐	☐	☐	☐	☐	☐	☐
Interested	☐	☐	☐	☐	☐	☐	☐
Frustrated	☐	☐	☐	☐	☐	☐	☐
Sad	☐	☐	☐	☐	☐	☐	☐

Yesterday, did you **spend time at home by yourself?** Without a spouse, partner, or anyone else present. *Tick one box.*
Yes ☐
No ☐ (skip next 2 questions)

How much time did you **spend at home by yourself yesterday?**
Hours___ Minutes___

How did you feel when you were **at home by yourself yesterday?** Rate each feeling on a scale from 0 (did not experience at all) to 6 (the feeling was extremely strong). *Tick one box on each line.*

I felt

	0	1	2	3	4	5	6
Happy	☐	☐	☐	☐	☐	☐	☐
Interested	☐	☐	☐	☐	☐	☐	☐
Frustrated	☐	☐	☐	☐	☐	☐	☐
Sad	☐	☐	☐	☐	☐	☐	☐

Additional module:

Overall, how did you feel **yesterday?** Rate each feeling on a scale from 0 (did not experience at all) to 6 (the feeling was extremely strong). *Tick one box on each line.*

I felt

	0	1	2	3	4	5	6
Happy	☐	☐	☐	☐	☐	☐	☐
Interested	☐	☐	☐	☐	☐	☐	☐
Frustrated	☐	☐	☐	☐	☐	☐	☐
Sad	☐	☐	☐	☐	☐	☐	☐
Enthusiastic	☐	☐	☐	☐	☐	☐	☐
Content	☐	☐	☐	☐	☐	☐	☐
Angry	☐	☐	☐	☐	☐	☐	☐
Tired	☐	☐	☐	☐	☐	☐	☐
Stressed	☐	☐	☐	☐	☐	☐	☐
Lonely	☐	☐	☐	☐	☐	☐	☐
Worried	☐	☐	☐	☐	☐	☐	☐
Bored	☐	☐	☐	☐	☐	☐	☐
Pain	☐	☐	☐	☐	☐	☐	☐
Depressed	☐	☐	☐	☐	☐	☐	☐
Joyful	☐	☐	☐	☐	☐	☐	☐

EXPERIENCED EMOTION QUESTIONS FROM THE GALLUP-HEALTHWAYS WELL-BEING INDEX

Did you experience **anger** during a lot of the day yesterday?
Yes ☐
No ☐

Did you experience **depression** during a lot of the day yesterday?
Yes ☐
No ☐

Did you experience **enjoyment** during a lot of the day yesterday?
Yes ☐
No ☐

Did you experience **happiness** during a lot of the day yesterday?
Yes ☐
No ☐

Did you experience **sadness** during a lot of the day yesterday?
Yes ☐
No ☐

Did you experience **stress** during a lot of the day yesterday?
Yes ☐
No ☐

Did you experience **worry** during a lot of the day yesterday?
Yes ☐
No ☐

Now, please think about yesterday, from the morning until the end of the day. Think about where you were, what you were doing, who you were with, and how you felt. **Did you learn or do something interesting yesterday?**
Yes ☐
No ☐

Now, please think about yesterday, from the morning until the end of the day. Think about where you were, what you were doing, who you were with, and how you felt. **Did you smile or laugh a lot yesterday?**
Yes ☐
No ☐

Now, please think about yesterday, from the morning until the end of the day. Think about where you were, what you were doing, who you were with, and how you felt. **Were you treated with respect all day yesterday?**
Yes ☐
No ☐

Now, please think about yesterday, from the morning until the end of the day. Think about where you were, what you were doing, who you were with, and how you felt. **Would you like to have more days just like yesterday?**
Yes ☐
No ☐

Additional module:

Did you experience **enthusiasm** during a lot of the day yesterday?
Yes ☐
No ☐

Did you experience **contentment** during a lot of the day yesterday?
Yes ☐
No ☐

Did you experience **frustration** during a lot of the day yesterday?
Yes ☐
No ☐

Did you experience **fatigue** during a lot of the day yesterday?
Yes ☐
No ☐

Did you experience **loneliness** during a lot of the day yesterday?
Yes ☐
No ☐

Did you experience **boredom** during a lot of the day yesterday?
Yes ☐
No ☐

Did you experience **pain** during a lot of the day yesterday?
Yes ☐
No ☐

What time did you wake up **yesterday**? _____

What time did you go to bed **yesterday**? _____

Did you feel well-rested **yesterday morning** (that is, you slept well the night before)? *Tick one box.*
Yes ☐
No ☐

Was **yesterday** a normal day for you or did something unusual happen?
Yes, just a normal day ☐
No, my day included unusual bad (stressful) things ☐
No, my day included unusual good things ☐

Intro: Please think about the **things you did yesterday**. How did you spend your time and how did you feel?

Yesterday, did you **watch TV**? *Tick one box.*
Yes ☐
No ☐ (skip next question)

How much time did you spend **watching TV yesterday**? *For example, if you spent one and a half hours, enter 1 for the hours and 30 for the minutes.*
Hours___ Minutes___

Yesterday, did you **work or volunteer**? *Tick one box.*
Yes ☐
No ☐ (skip next question)

How much time did you spend **working or volunteering yesterday**? *For example, if you spent nine and a half hours, enter 9 for the hours and 30 for the minutes.*
Hours___ Minutes___

Yesterday, did you go for a **walk or exercise**? *Tick one box.*
Yes ☐
No ☐ (skip next question)

How much time did you spend **walking or exercising yesterday**? *For example, if you spent 30 minutes, enter 0 for the hours box and 30 for the minutes.*
Hours___ Minutes___

Yesterday, did you do any **health-related activities other than walking or exercise**? *For example, visit a doctor, take medications, or have a treatment. Tick one box.*
Yes ☐
No ☐ (skip next question)

How much time did you spend doing **health-related activities yesterday?**
Hours___ Minutes___

Yesterday, did you **travel or commute**? *For example, by car, train, bus, etc. Tick one box.*
Yes ☐
No ☐ (skip next question)

APPENDIX A   *147*

How much time did you spend **traveling or commuting yesterday**?
Hours___ Minutes___

Yesterday, did you **spend time with friends or family**? *Tick one box.*
Yes ☐
No   ☐ (skip next question)

How much time did you spend **with friends or family yesterday**?
Hours___ Minutes___

Yesterday, did you **spend time at home by yourself?** Without a spouse, partner, or anyone else present. *Tick one box.*
Yes ☐
No   ☐ (skip next question)

How much time did you **spend at home by yourself yesterday**?
Hours___ Minutes___

How did you feel when you were **walking or exercising**? Rate each feeling on a scale from 0 (did not experience at all) to 6 (the feeling was extremely strong). *Tick one box on each line.*

I felt

	0	1	2	3	4	5	6
Happy	☐	☐	☐	☐	☐	☐	☐
Interested	☐	☐	☐	☐	☐	☐	☐
Frustrated	☐	☐	☐	☐	☐	☐	☐
Sad	☐	☐	☐	☐	☐	☐	☐

ExWB QUESTIONNAIRE FROM THE HWB-12 MODULE

SOURCE: Smith and Stone (2011).

Now we would like you to think about yesterday. What did you do yesterday and how did you feel?
To begin, please tell me what time you woke up **yesterday**. _____
And what time did you go to sleep **yesterday**? _____
Now please take a few quiet seconds to recall your activities and experiences **yesterday**.

Good, now I have questions about your experiences **yesterday**.

[Randomize order of emotions]

Yesterday, did you feel **happy**? Would you say
Not at all, A little, Somewhat, Quite a bit, Very
☐ ☐ ☐ ☐ ☐

Yesterday, did you feel **enthusiastic**? Would you say
Not at all, A little, Somewhat, Quite a bit, Very
☐ ☐ ☐ ☐ ☐

Yesterday, did you feel **content**? Would you say
Not at all, A little, Somewhat, Quite a bit, Very
☐ ☐ ☐ ☐ ☐

Yesterday, did you feel **angry**? Would you say
Not at all, A little, Somewhat, Quite a bit, Very
☐ ☐ ☐ ☐ ☐

Yesterday, did you feel **frustrated**? Would you say
Not at all, A little, Somewhat, Quite a bit, Very
☐ ☐ ☐ ☐ ☐

Yesterday, did you feel **tired**? Would you say
Not at all, A little, Somewhat, Quite a bit, Very
☐ ☐ ☐ ☐ ☐

Yesterday, did you feel **sad**? Would you say
Not at all, A little, Somewhat, Quite a bit, Very
☐ ☐ ☐ ☐ ☐

Yesterday, did you feel **stressed**? Would you say
Not at all, A little, Somewhat, Quite a bit, Very
☐ ☐ ☐ ☐ ☐

Yesterday, did you feel **lonely**? Would you say
Not at all, A little, Somewhat, Quite a bit, Very
☐ ☐ ☐ ☐ ☐

Yesterday, did you feel **worried**? Would you say
Not at all, A little, Somewhat, Quite a bit, Very
☐ ☐ ☐ ☐ ☐

Yesterday, did you feel **bored**? Would you say
Not at all, A little, Somewhat, Quite a bit, Very
☐ ☐ ☐ ☐ ☐

Yesterday, did you feel **pain**? Would you say
Not at all, A little, Somewhat, Quite a bit, Very
☐ ☐ ☐ ☐ ☐

Additional module: [*Randomize order of emotions*]

Yesterday, did you feel **depressed**? Would you say
Not at all, A little, Somewhat, Quite a bit, Very
☐ ☐ ☐ ☐ ☐

Yesterday, did you feel **joyful**? Would you say
Not at all, A little, Somewhat, Quite a bit, Very
☐ ☐ ☐ ☐ ☐

Yesterday, did you **learn or do something interesting**? Would you say
Not at all, A little, Somewhat, Quite a bit, Very
☐ ☐ ☐ ☐ ☐

Did you feel **well-rested yesterday morning** (that is, you slept well the night before)?
Yes ☐ No ☐

Was **yesterday** a normal day for you or did something unusual happen? *Tick one box.*
Yes, just a normal day ☐
No, my day included unusual bad (stressful) things ☐
No, my day included unusual good things ☐

Intro: Please think about the **things you did yesterday**. How did you spend your time and how did you feel?

Yesterday, did you **watch TV**? *Tick one box.*
Yes ☐
No ☐ (**skip next question**)

How much time did you spend **watching TV yesterday**? *For example, if you spent one and a half hours, enter 1 for the hours and 30 for the minutes.*
Hours___ Minutes___

Yesterday, did you **work or volunteer?** *Tick one box.*
Yes ☐
No ☐ (skip next question)

How much time did you spend **working or volunteering yesterday?** *For example, if you spent nine and a half hours, enter 9 for the hours and 30 for the minutes.*
Hours___ Minutes___

Yesterday, did you go for a **walk or exercise?** *Tick one box.*
Yes ☐
No ☐ (skip next question)

How much time did you spend **walking or exercising yesterday?** *For example, if you spent 30 minutes, enter 0 for the hours box and 30 for the minutes.*
Hours___ Minutes___

Yesterday did you do any **health-related activities other than walking or exercise?** *For example, visit a doctor, take medications, or have a treatment. Tick one box.*
Yes ☐
No ☐ (skip next question)

How much time did you spend doing **health-related activities yesterday?**
Hours___ Minutes___

Yesterday did you **travel or commute?** *For example, by car, train, bus, etc. Tick one box.*
Yes ☐
No ☐ (skip next question)

How much time did spend **traveling or commuting yesterday?**
Hours___ Minutes___

Yesterday did you **spend time with friends or family?** *Tick one box.*
Yes ☐
No ☐ (skip next question)

How much time did you spend **with friends or family yesterday?**
Hours___ Minutes___

Yesterday, did you **spend time at home by yourself?** Without a spouse, partner, or anyone else present. *Tick one box.*
Yes ☐
No ☐ (skip next question)

How much time did you **spend at home by yourself yesterday?**
Hours___ Minutes___

How did you feel when you were **walking or exercising?** Rate each feeling on a scale from 0 (did not experience at all) to 6 (the feeling was extremely strong). *Tick one box on each line.*

I felt

	0	1	2	3	4	5	6
Happy	☐	☐	☐	☐	☐	☐	☐
Interested	☐	☐	☐	☐	☐	☐	☐
Frustrated	☐	☐	☐	☐	☐	☐	☐
Sad	☐	☐	☐	☐	☐	☐	☐

Appendix B

The Subjective Well-Being Module of the American Time Use Survey: Assessment for Its Continuation

NOTE: For brevity, several pages of front matter and the appendix of Biographical Sketches of Panel Members that appear in the published version of this interim report have been omitted in this version. The reference list for citations in this appendix is at the end of the appendix.

The published report is available from the National Academies Press at http://www.nap.edu/catalog.php?record_id=13535.

Erratum: The citations and reference item given as Boeham and Kobzansky, (2012) should be Boehm and Kobzansky (2012).

APPENDIX B

The Subjective Well-Being Module of the American Time Use Survey: Assessment for Its Continuation

Panel on Measuring Subjective Well-Being in a Policy-Relevant Framework

Committee on National Statistics

Division of Behavioral and Social Sciences and Education

NATIONAL RESEARCH COUNCIL
OF THE NATIONAL ACADEMIES

THE NATIONAL ACADEMIES PRESS
Washington, D.C.
www.nap.edu

PANEL ON MEASURING SUBJECTIVE WELL-BEING IN A POLICY-RELEVANT FRAMEWORK

ARTHUR A. STONE (*Chair*), Department of Psychiatry and Behavioral Sciences, Stony Brook University
NORMAN M. BRADBURN, Department of Psychology, University of Chicago
LAURA L. CARSTENSEN, Department of Psychology, Stanford University
EDWARD F. DIENER, Department of Psychology, University of Illinois at Urbana-Champaign
PAUL H. DOLAN, Department of Social Policy, London School of Economics and Political Science
CAROL L. GRAHAM, The Brookings Institution, Washington, DC
V. JOSEPH HOTZ, Department of Economics, Duke University
DANIEL KAHNEMAN, Woodrow Wilson School, Princeton University
ARIE KAPTEYN, The RAND Corporation, Santa Monica, CA
AMANDA SACKER, Institute for Social and Economic Research, University of Essex, United Kingdom
NORBERT SCHWARZ, Department of Psychology, University of Michigan
JUSTIN WOLFERS, The Wharton School, University of Pennsylvania

CHRISTOPHER MACKIE, *Study Director*
ANTHONY S. MANN, *Program Associate*

Contents

SUMMARY

1 BACKGROUND AND OVERVIEW
   1.1 Structure and Content of ATUS and the SWB Module
   1.2 Objectives of the SWB Module
   1.3 Uses of Data on Subjective Well-Being

2 ONGOING AND POTENTIAL RESEARCH APPLICATIONS
   2.1 Time Use, Emotional Well-Being, and Unemployment
   2.2 Assessing Validity of Short Versions of the Day Reconstruction Method
   2.3 Episode-Based Pain Studies
   2.4 End-of-Life Care
   2.5 Transportation

3 ASSESSMENT
   3.1 Value of the SWB Module Data to Date
   3.2 Cost of Discontinuing the Module
   3.3 Value of a Third Wave

REFERENCES

Summary

The American Time Use Survey (ATUS), conducted by the Bureau of Labor Statistics, included a Subjective Well-Being (SWB) module in 2010 and 2012; the module, funded by the National Institute on Aging (NIA), is being considered for inclusion in the ATUS for 2013. The National Research Council (NRC) was asked to evaluate measures of self-reported well-being and offer guidance about their adoption in official government surveys. The charge for the study included an interim report to consider the usefulness of the ATUS SWB module and specifically the value of continuing it for at least one more wave. Among the key points raised in this report are the following:

- *Value* The ATUS SWB module is the only federal government data source of its kind—linking self-reported information on individuals' well-being to their activities and time use. Important research has already been conducted using the data (for example, on the effects of unemployment and job search on people's self-reported well-being), and work conducted with other, similar data sets has indicated the potential of the module to contribute to knowledge that could inform policies in such areas as health care and transportation. While the NRC Panel has not yet concluded its assessment of the policy usefulness of including one or more kinds of self-reported well-being measures on a regular basis in government surveys, it sees a value to continuing the ATUS SWB module in 2013. Not only will another year of data support research, but

it will also provide additional information to help refine any SWB measurements that may be added to ongoing official statistics.
- *Methodological Benefits* A third wave of data collection will enlarge samples by pooling data across years, which will enable more detailed study and comparison than has been possible to date of population subgroups, such as people in a given region and specific demographic groups (e.g., young people, the elderly). Because two new questions—one on overall life satisfaction and one on whether respondents' reported emotional experiences yesterday were "typical"—were introduced to the module only in 2012, at least one additional wave of the survey is needed to assess changes in responses to those questions over time.
- *Cost and Effects on the ATUS* As a supplement to an existing survey, the marginal cost of the module, which adds about 5 minutes to the ATUS, is small. While further study of the module's effects on response and bias in the main ATUS should be undertaken, it appears likely that these effects are modest because the module comes at the end of the survey after people have already been asked to report their activities for the preceding day.
- *New Opportunities* A third wave of the survey could also be used for experiments to improve the survey structure, should the module become permanent. The ATUS SWB module could be the basis for a standardized set of questions that could be added to other surveys which, together, might provide useful information about the causes and consequences of self-reported well-being in the general population.

1

Background and Overview[1]

Research on subjective or self-reported well-being (SWB) has been ongoing for several decades, with the past few years seeing an increased interest by some countries in using SWB measures to evaluate government policies and provide a broader assessment of the health of a society than is provided by such standard economic measures as Gross Domestic Product (see, for example, Stiglitz, Sen, and Fitoussi, 2009). The National Institute on Aging (NIA) and the United Kingdom Economic and Social Research Council asked a panel of the National Research Council's Committee on National Statistics to review the current state of research knowledge and evaluate methods for measuring self-reported well-being and to offer guidance about adopting SWB measures in official population surveys (see Box 1-1 for the full charge to the panel). NIA also asked the panel to prepare an interim report on the usefulness of the Subjective Well-Being module of the American Time Use Survey (ATUS), with a view as to the utility of continuing the module in 2013.

The SWB module is the only national data source in the United States that links self-reported well-being information to individuals' activities and time-use patterns. It provides researchers with unique insights that are only revealed by melding ratings of affect with time use information. The SWB module, overseen by the Bureau of Labor Statistics (BLS) and sponsored by

---

[1] This section draws heavily from a presentation to the panel by Rachel Kranz-Kent of BLS, and from the *Federal Register*, Volume 76, Number 134 (July 13, 2011): http://webapps.dol.gov/federalregister/HtmlDisplay.aspx?DocId=25169&AgencyId=6&DocumentType=3 (accessed August 24, 2012).

APPENDIX B

> **BOX 1-1**
> **Panel Charge**
>
> An ad hoc panel will review the current state of research and evaluate methods for the measurement of subjective well-being (SWB) in population surveys. On the basis of this evaluation, the panel will offer guidance about adopting SWB measures in official government surveys to inform social and economic policies. The study will be carried out in two phases. The first phase, which is the subject of this statement of task, is to consider whether research has advanced to a point that warrants the federal government collecting data that allow aspects of the population's SWB to be tracked and associated with changing conditions. The study will focus on experienced well-being (e.g., reports of momentary positive and rewarding, or negative and distressing, states) and time-based approaches (some of the most promising of which are oriented toward monitoring misery and pain as opposed to "happiness"), though their connection with life-evaluative measures will also be considered. Although primarily focused on SWB measures for inclusion in U.S. government surveys, the panel will also consider inclusion of SWB measures in surveys in the United Kingdom and European Union, in order to facilitate cross-national comparisons in addition to comparisons over time and for population groups within the United States. The panel will prepare a short interim report on the usefulness of the American Time Use Survey subjective well-being module, and a final report identifying potential indicators and offering recommendations for their measurement. A later, separate second phase will seek to develop a framework modeled on the National Income and Product Accounts to integrate time-based inputs and outputs, and SWB measures, into selected satellite, or experimental, subaccounts.

NIA, was developed with guidance from several noted academics—Angus Deaton, Daniel Kahneman, Alan Krueger, David Schkade, and Arthur Stone among them—working in the field.

Though the SWB module has only been in existence since 2010, it is not too early to begin assessing its potential value to researchers and policy makers. The purpose of this report is to inform planning discussions about the module's future—it discusses the costs and benefits of a third wave of data collection, whether the survey module should be modified, and whether experiments should be done to improve the module should it become permanent.

This brief report is intended to fulfill only one narrow aspect of the panel's broader task as described in Box 1-1. It provides (1) an overview of the ATUS and the SWB module; (2) a brief discussion of research applications to date; and (3) preliminary assessment of the value of SWB module data. The panel's final report will address issues of whether research has advanced to the point that SWB measures—and which kinds of measures—should be regularly included in major surveys of official statistical agencies to help inform government economic and social policies.

1.1 STRUCTURE AND CONTENT OF ATUS AND THE SWB MODULE

The ATUS is the first federally administered, continuous survey on time use in the United States (and in the world). It is designed to obtain estimates of the time spent by respondents in childcare, at work, traveling, sleeping, volunteering, engaged in leisure pursuits, and a wide range of other activities. Time-use data augment income and wage data for individuals and families that analysts can use to create a more complete picture of quality of life in a society. Along with income and product data, information about time-use patterns is essential for research that evaluates the contribution of nonmarket work to national economies. The data also enable comparisons between nations that have different mixes of market and nonmarket production modes. To illustrate, the households of two countries may enjoy similar home services and amenities—quality of meals, level of home cleaning and maintenance, elder and child care, etc.—but one may perform more of these tasks themselves (home production) while the other may more typically hire the tasks out in the market. The latter economy will register higher per capita gross domestic product even though the standard of living may be comparable in the two countries. Relatedly, countries may vary in the amount of time that individuals must work to achieve a given material standard of living, resulting in different amounts of leisure. This difference would also not show up directly in market (only) measures of economic activity, yet it is likely that it affects well-being.

The ATUS provides nationally representative estimates of how people spend their time. It has been conducted continuously since 2003. The survey sample is a repeated cross-section of individuals who are drawn from U.S. households completing their eighth and final month of interviews for the Current Population Survey (CPS). One individual from each household is selected to take part in one computer-assisted telephone interview. Respondents are interviewed for the ATUS between two and five months after they rotate out of the CPS.

Interviewers ask respondents to report all of their activities for one specified 24-hour day, the day prior to the interview. Respondents also report who was with them during activities, where they were, how long each activity lasted, and if they were paid. For the ATUS (following the core time diary questions but prior to the SWB module) some of the CPS information—for example, about who is living in the household and labor

APPENDIX B

force status—is confirmed and updated.[2] Measurement of socioeconomic well-being based on the ATUS is enhanced by its connection to the CPS which is rich in socio-demographic variables—namely, characteristics of the individual and the household including labor force status, income, state of residence, educational attainment, race and ethnicity, nativity, detailed marital status (divorced, never married, etc.), and disability status.[3]

The SWB module adds to the substantive content of the ATUS by revealing not only what people are doing with their time, but also how they experience their time—specifically how happy, tired, sad, stressed, and in pain they felt while engaged in specific activities on the day prior to the interview.[4] This information has numerous practical applications for sociologists, economists, educators, government policy makers, businesspersons, health researchers, and others. The module follows directly after the core ATUS; it was administered on an ongoing basis during 2010 and is being done again during 2012. The module surveys individuals aged 15 and over from a nationally representative sample of approximately 2,190 households each month.

Respondents are asked questions about three activities selected with equal probability from those reported in the ATUS time diary (the well-being module questions are asked immediately after the core ATUS) (see Box 1-2). A few activities—sleeping, grooming, and private activities—are never included in the SWB module. The time diary refers to the core part of the ATUS, in which respondents report the activities they did from 4 a.m. on the day before the interview to 4 a.m. on the day of the interview. The precodes listed in Box 1-2 are for activities that are straightforward to code, but they are in no way representative of the full activity lexicon used by ATUS coders. The vast majority of ATUS activities are typed into the collection instrument (verbatim) and then coded in a separate processing step.[5] The module also collects data on whether respondents were interacting with anyone while doing the selected activities and how meaningful the activities were to them.

Respondents are asked to rate, for each of the three randomly selected activities, six feelings—pain, happy, tired, sad, stressed, and meaningful—

---

[2] Technical details of the sample design and the survey methodology can be found in the *American Time Use Survey User's Guide: Understanding ATUS 2003-2011*. Available: http://www.bls.gov/tus/atususersguide.pdf [September 3, 2012].

[3] Information about who is living in the household and about labor force status is updated in the ATUS, which is important since the CPS data are a little dated by the time the ATUS interview takes place.

[4] The module questionnaire can be found at http://www.bls.gov/tus/wbmquestionnaire.pdf [August 2012].

[5] There are more than 400 possible activity codes; a full list can be found at http://www.bls.gov/tus/lexiconnoex2011.pdf [June 27, 2012].

> **BOX 1-2**
> **ATUS Question Identifying an Activity**
>
> So let's begin. Yesterday, Monday, at 4:00 a.m., what were you doing?
>
> - Use the slash key (/) for recording separate/simultaneous activities.
> - Do not use precodes for secondary activities.
>
> 1. Sleeping
> 2. Grooming (self)
> 3. Watching TV
> 4. Working at main job
> 5. Working at other job
> 6. Preparing meals or snacks
> 7. Eating and drinking
> 8. Cleaning kitchen
> 9. Laundry
> 10. Grocery shopping
> 11. Attending religious service
> 12. Paying household bills
> 13. Caring for animals and pets
> 14. Don't know/Can't remember
> 15. Refusal/None of your business

on a scale from 0 to 6: 0 means the feeling was not present, and 6 means the feeling was very strong (see Box 1-3).

The following health related questions (paraphrased here) are also asked after the three random activity episodes are chosen:

- Did you take pain medication yesterday?
- When you woke up yesterday, how well rested did you feel?
- Do you have hypertension?
- Would you say your health in general is excellent, very good, good, fair, or poor?

This information creates opportunities to analyze interactions between health states and reported assessments of emotional states. This is important because daily experience is linked to health status and other outcomes via channels such as worry and stress on the one hand, and pleasure and enjoyment on the other.

> **BOX 1-3**
> **ATUS SWB Text Asking Respondents to Rate Strength of Feeling During Specific Activities**
>
> Between 12:00 p.m. and 1:00 p.m. yesterday, you said you were eating and drinking. The next set of questions asks how you felt during that particular time.
>
> Please use a scale from 0 to 6, where 0 means you did not otherwise experience this feeling at all and a 6 means the feeling was very strong. You may choose any number 0, 1, 2, 3, 4, 5, or 6 to reflect how strongly you experienced this feeling during this time.

1.2 OBJECTIVES OF THE SWB MODULE

The ATUS SWB module was initially designed to collect information primarily on experienced ("hedonic") well-being—that is, about people's emotions associated with a recent time period and the activities that occurred during that period. The hedonic dimension of well-being is directly related to the environment or context in which people live—the quality of their jobs, their immediate state of health, the nature of their commute to work, and the nature of their social networks—and is reflected in positive and negative affective states. These kinds of hedonic measures contrast with self-reported assessments of overall life satisfaction or happiness. Such "evaluative" well-being measures are more likely to reflect people's attitudes about their lives as a whole.

The first, 2010, module included only hedonic measures. The second wave (conducted in 2012) includes two additional questions, one on overall life satisfaction and one on whether or not recent emotional experience was typical. The life satisfaction responses are collected using the Cantril ladder scale.[6] As noted on the BLS supporting statement for the project (p. 2), asking the Cantril ladder question enables researchers "to build a link between time use and day reconstruction methods of measuring well-being on the one hand, and standard life evaluation questions on the other . . . a direction of research that has not been possible to date." The life

---

[6]The Cantril Self-Anchoring Scale asks respondents to imagine a ladder with steps numbered from 0 at the bottom to 10 at the top, in which the top of the ladder represents the best possible life for them and the bottom of the ladder represents the worst possible life. They are asked which step of the ladder they personally feel they stand on at this time (for a present assessment). For a good description and discussion of the Cantril Scale, see Diener et al. (2009).

evaluation question enhances the value both of the ATUS supplement and other surveys that use a Cantril ladder question.

Measurement of both *experienced* well-being (i.e., reports of momentary positive and rewarding or negative and distressing states) and *evaluative* well-being (i.e., cognitive judgments of overall life satisfaction or dissatisfaction) extends the policy value of the SWB module data. The value added comes from what can be learned from differences between what the two measures show. For example Kahneman and Deaton (2010, p. 1) find that "emotional well being and life evaluation have different correlates in the circumstances of people's lives" and particularly striking "differences in the relationship of these aspects of well being to income."

Distinguishing between different dimensions of well-being also allows investigation of psychological changes associated with aging (e.g., reduced mobility) that might affect both these dimensions of well-being. Another area where the two dimensions provide complementary information is job satisfaction. Getting promoted or obtaining a new job that entails long hours might raise a worker's evaluative well-being, but the associated stress might reduce experienced well-being, at least in the short term. Similar comparisons could be made across professions. Respondents' reported differences between experience and evaluative measures might also help explain why some people attach high meaning to work, career, and related time commitments while others focus more on simple day-to-day contentment and how or if these correlations vary across age, income, and other demographic or cohort factors. For education research, measures of multiple dimensions of subjective well-being may help provide an understanding of why students make (or do not make) the investments in schooling choices that they do (or do not) make.

The second new question for 2012 asks whether the respondents' emotional experience yesterday (the day before the interview) was typical for that day of the week:

> Thinking about yesterday as a whole, how would you say your feelings, both good and bad, compared to a typical Monday? Were they better than a typical Monday, the same as a typical Monday, or worse than a typical Monday (respondents answer "better," "the same," or "worse").

This question may provide insights about day of week effects and day to day variation in reported well-being scores.

1.3 USES OF DATA ON SUBJECTIVE WELL-BEING

Data from the SWB module supports the BLS mission of providing relevant information on economic and social issues. The data provide a richer description of work experience; specifically, these data describe how individuals feel (tired, stressed, in pain) during work episodes compared to non-work episodes, and how often workers interact on the job. Data from the module can also be used to measure whether the amount of physical pain that workers experience varies by occupation and disability status. The fact the SWB module can be linked to demographic characteristics of respondents—labor force status, occupation, earnings, household composition, school enrollment status, and other characteristics captured on the core ATUS and CPS—opens up a wide array of possible studies on the correlates of self-reported well-being.[7]

Collection of data on subjective well-being also supports the mission of the module's sponsor, NIA, to improve the health and well-being of older Americans. Examples of questions that can be answered include:

Do older workers experience more pain than younger workers on and off the job?

Is the age-pain gradient related to differences in activities or differences in the amount of pain experienced during a given set of activities?

Do those in poor health spend time in different activities relative to those in good health?

To date, much of the research on nonmarket components of health and well-being has been informed by global assessments of positive or negative affect averaged over time that are divorced from measures of time use or context. Nor has that research typically addressed age differences or age-related changes in these associations. In this vein, data from the SWB module might inform policies on redesigning cities to support healthy aging, the allocation of funds to programs that affect older populations, and changes to the health care system to support better maintenance of good health. Researchers have already begun to explore which aspects of experienced and evaluative well-being, time use, and context promote or impede healthy aging. Further work can be done to examine the unique correlative and predictive associations of evaluated and experienced well-being with health and with differences related to life stage, retirement status, and individual characteristics.

---

[7] In addition, because the ATUS is conducted through the year, it is possible to study seasonal effects on well-being—a topic of interest in a number of research areas.

2

Ongoing and Potential Research Applications

Compelling evidence indicates that higher levels of subjective or self-reported well-being (SWB) are associated with a range of desirable outcomes, from better health and greater longevity to stable social relationships and even to economic productivity. Daily stress, for example, has been shown to correlate quite strongly with illness, and higher levels of hedonic well-being (positive feelings) with lower incidence of cardiovascular disease (Boeham and Kubzansky, 2012; Huppert, 2009). Based on the current evidence, generated from research using a variety of methods, one could even reasonably conclude that SWB is likely a causal factor for some health outcomes. This in itself is a compelling reason to gather data on and analyze the subjective-well-being of the population.

Though data from the 2010 ATUS SWB module have only been publicly available since November 2011 (2012 data will not be available until next year), research using those data is already emerging. This section identifies some of that work to provide a sense of the range of applications.

2.1 TIME USE, EMOTIONAL WELL-BEING, AND UNEMPLOYMENT

In an analysis of the differences in time use and emotional well-being between employed and unemployed people—for specific activities identified using the ATUS sample—Krueger and Mueller (2012) show that the unemployed get less enjoyment out of leisure and report higher levels of sadness during specific activities relative to employed (the sadness decreases abruptly at the time of employment).[8] This study leans more heavily on data from the Survey of Unemployed Workers in New Jersey since its longitudinal structure, in contrast to the repeated cross-sectional measurement in ATUS, allows consideration of fixed effects—that is, to look at *within* group variation—but is indicative of the importance of being able to link data on subjective well-being to specific events.

2.2 ASSESSING VALIDITY OF SHORT VERSIONS OF THE DAY RECONSTRUCTION METHOD (DRM)

Vicki Freedman, Richard Gonzalez, Lindsay Ryan, Norbert Schwarz, Jacqui Smith, and Robert Stawski, are comparing DRM—which involves asking respondents to reconstruct and describe episodes of the previous day and the feelings they experienced during each—with shorter survey approaches that retain a subset of DRM features.[9] This work is comparing findings from the Health and Retirement Study (HRS) with DRM data collected in the Panel Study of Income Dynamics (PSID), the ATUS SWB module, and the American Life Panel.[10] The minimum features necessary for a short, reliable, and valid survey index of experienced well-being are unknown, though the target length of a survey measure being tested in their study is 3-5 minutes.

This kind of evaluation is central to determining how broadly subjective measures can potentially be integrated into policy analyses and national statistics. Adding a standardized module of well-being questions to surveys covering a wide range of domains (health, employment, etc.) is necessary for understanding covariates of (and developing statistics on) population well-being. However, such an integrated strategy will only be feasible if the modules are minimally burdensome and retain validity across contexts and if the short-version questionnaires are sufficiently robust in the information they produce.

---

[8] More generally, the ATUS SWB module has the potential to add richness to research on trends in leisure and leisure inequality (see, e.g., Aguiar and Hurst, 2007) and on the link between leisure and well-being (see, Meyer and Sullivan, 2009, which examines changes in the distribution of well-being as a function of not just consumption of goods and services, but also consumption of time, by incorporating information based on self-reported measures.

[9] A brief description of this research in progress can be found at http://micda.psc.isr.umich.edu/project/detail/35382 (accessed July 17, 2012).

[10] One appealing argument for collecting time-use and hedonic data through an approach like that of the day reconstruction method is that it can then be used to compute other measures of experienced well-being such as the U-index, which measures the proportion of time individuals spend in an "unpleasant," "undesirable," or "unhappy" state (see Krueger and Stone, 2008). A focus on the U-index would be justified if policy makers want to pay attention to the incidence of negative feelings and their health and other consequences.

2.3 EPISODE-BASED PAIN STUDIES

Two additional sets of analyses that use ATUS or ATUS-like data are worth noting because they provide an indication of potential uses of data from the SWB module. In a recent study, Krueger and Stone (2008) measured pain during specific random periods of time, which allowed them to study how reported (recalled) levels of pain affected activities of daily living in particular segments of the sample population. This approach is novel relative to the global assessment methodologies typically used in population studies. The authors used data from the Princeton Affect and Time Survey (PATS), which employs a similar data collection methodology and the same general procedures as ATUS: "yesterday" is reconstructed through computer-assisted telephone interviews, and then three episodes from those identified are randomly drawn and information is collected about affect and pain.

Similar studies could be done even more robustly using ATUS, as PATS allowed only 3,982 respondents, while there were more than 12,000 in the 2010 ATUS sample. In addition, the PATS sample was likely less representative than the ATUS sample. Even with these limitations in PATS (relative to ATUS), the finding from this study were clear and robust: one was that those with lower income or less education reported higher average pain than did those with higher income or more education, and another was that average pain ratings reached a plateau between the ages of about 45 years and 75 years. The results of this study suggest even greater potential for the value of ATUS for pain studies—an area where there is an increasing demand for research.

Stone and Deaton have recently begun work, using the 2010 SWB module data, to examine the hypothesis that people with different employment status (working/nonworking) and occupations (using standard labor categories) experience different levels of pain throughout the day—and not just on the job.[11] Possible explanations for variation in reported pain levels include the differing physical demands of different occupations; these pain-occupation relationships may vary by age or gender. The researchers first examined pain, rated on a scale from 0 (did not feel any pain) to 6 (severe pain), for a broad employment status variable. They found those who were employed had less pain than those who were unemployed and were looking for work or who were retired or disabled. People in management, business, and financial occupations had lower pain levels than almost all of the other occupational categories (controlling for age and sex). People in occupations that are judged as having higher levels of manual labor also

---

[11] This work is being done by Arthur Stone (Stony Brook University) and Angus Deaton (Princeton University).

reported more daily pain. Pain was also higher on average during times respondents reported being at work in comparison with other activities. Other aspects of hedonic well-being—e.g., specific emotions, such as stress or enjoyment—may ultimately be examined in much the same way.

Similarly, it is possible to test if pain was higher at work or during periods not at work, and whether or not this distinction interacted with type of occupation: Do those with physically demanding jobs experience more pain on the job than when not working? Is this pattern less pronounced for less physically demanding occupations? These analyses have begun to reveal the capability of the detailed, daily data of the ATUS to address both between- and within-subjects questions, and highlight the richness of the data.

2.4 END-OF-LIFE CARE

Various well-being measures have been used for some time to supplement measures of objective health in clinical and epidemiological research, particularly by those interested in broadening the concept of health beyond the absence of illness to include the presence of positive health, functioning, and other quality-of-life dimensions.

Policies oriented toward improving care for the chronically ill or for end-of-life care, for example, could benefit from better data on the impact that various treatments have on patients and on their families and careers. Data on subjective well-being could be useful in this area, especially for monitoring those who are providing care, such as family members. The data could identify where targeted studies are needed, such as when quality is at least as important as quantity of life. The distinction between hedonic well-being and other dimensions of well-being addressed in the 2012 SWB module may be especially important for the end of life, when the balance between predominantly purposeful and pleasurable activities might change.

In addition, the well-being of eldercare providers is of interest to policy makers because the elderly population is growing, along with a reliance on informal care providers to assist them. Researchers may be able to take advantage of a change that was made to the ATUS in 2011, when questions that identify eldercare providers and eldercare activities were added.

2.5 TRANSPORTATION

Transportation has been identified as a potentially key determinant in the quality of people's lives. For example, when the transportation infrastructure is of poor quality or overcrowded, congestion and unreliable travel times inhibit the ability of individuals to engage in enjoyable or productive activities. Therefore, modeling the relationship between travel behavior and activities with measures of well-being represents a potential policy application of time use and well-being data (Diener, 2006; Steg and Gifford, 2005). Archer et al. (2012, p. 1) describe how transportation forecasting models may be used to help inform policy and investment decisions; they use the 2010 ATUS and SWB module data to develop a multivariate model designed to "capture the influence of activity-travel characteristics on subjective well-being while accounting for unobserved individual traits and attitudes that predispose people when it comes to their emotional feelings." They find that "activity duration, activity start time, and child accompaniment significantly impact feelings of well-being for different activities" (including travel). The authors add that "by integrating the well-being model presented in this paper with activity-based microsimulation models of travel demand, measures of well-being for different demographic segments may be estimated and the impacts of alternative policy and investment decisions on quality of life can be better assessed."

3

Assessment

3.1 VALUE OF THE SWB MODULE DATA TO DATE

It is still early to gauge the research and policy value of data emerging from the ATUS SWB module. Even so, the kinds of research described above provide a preliminary indication of the insights that can be drawn from the ability to combine time-use information (as it links to specific activities) and self-assessments of well-being during those periods, which have relevance to policies ranging from commuting and home production to eldercare and maintaining good health. Without established and consistent historical data that combine time use and emotional experience, researchers would be limited to analyzing trends in evaluated time use that are difficult to tie to specific determinants.

Several characteristics of the SWB module data contribute to its value:

- Its status as the only national data source on subjective well-being that is linked to activities and time use.
- Its Day Reconstruction Method (DRM)-like capability, unavailable with most other data sources on subjective well-being.
- Its large enough sample sizes (especially if pooled over multiple survey years) to accommodate analyses of important subgroups of the population.
- Its ability to facilitate research to begin solving difficult measurement and conceptual issues that have historically plagued work on subjective well-being.

The fact that the ATUS SWB module is the only federal government data source of its kind gives it a potentially very high value. In particular, its approximation of the DRM is unique.[12] As described above, linking of emotional states to daily experience may be the most directly relevant dimension of subjective well-being to policy. It is important to know how people feel when they are working, commuting, taking care of the old and the young, etc. In addition, identifying the context in which such activities take place, and asking respondents to rate well-being in that context (in the case of the ATUS, of the previous day) has the advantage of eliciting specific memories and, in turn, reducing bias associated with respondent recall.

More generally, there has been enough progress in research on the measurement of subjective well-being to pinpoint specific policy domains and questions for which such data are useful. For example, cross-sectional data have proven important for research assessing the relative impact on people of income and unemployment[13] and marriage and marital dissolution (Deaton, 2011, p. 50) and, more generally, on the effect of policies where large nonmarket components are involved (e.g., standard of living during end-of-life medical treatment). Data on subjective well-being have the potential to augment information in any situation in which market data are unavailable or not relevant and policy makers require criteria for choosing one course of action among two or more alternatives. In these cases, a range of evidence—revealed preference, stated preference, and subjective well-being measures—can usefully be drawn upon. And well-being measures that are tied to specific activities add a great deal of subtlety to these analysis; for example, while perhaps unemployed persons are able to engage more in activities they like to do (spend time with friends or relatives, rest, watch

---

[12] The day reconstruction method is itself an approximation of more time-consuming experience sampling and ecological momentary assessment methods; however, the day reconstruction method captures information about *episodes* while the ecological momentary assessment method typically captures information about *moments* (Christodoulou, Schneider, and Stone, 2012). Simplified versions of the experience sampling and ecological momentary assessment methods—which, in some, sense represent the gold standard since they involve repeated assessment in real time of people's current hedonic well-being—are necessitated by burden, time, and intrusiveness constraints in surveys. Though research is under way on the issue, it is still an open question how well, and under what conditions, the day reconstruction method approximation is adequate and useful.

[13] One could reasonably conclude that addressing the recent high rate of unemployment was made even more urgent by findings from research on subjective well-being showing that, in terms of individuals' utility, more was involved than simply an income effect. As Krueger and Mueller (2012) note, unemployment takes an emotional toll on people even while they are engaged in leisure activities. This calls into question an earlier conclusion by economists that people's decreases in well-being because of unemployment may be partially compensated by increases in leisure.

television, etc.), perhaps they enjoy each of those activities less relative to the employed.

It will be a task for this Panel's final report to provide an assessment of the extent to which subjective measures—including both global, evaluative measures and the more experiential measures that are the focus of this module—can or should be used to guide policy. Collecting data within the context of the ATUS has the potential to help researchers and policy makers evaluate whether these measures can be used in this way.

3.2 COST OF DISCONTINUING THE MODULE

The cost of discontinuing the module could be large since—if the value of such data became more apparent at some point in the future—restarting the survey would likely entail repeating start-up tasks and drawing again on political capital to make it happen. More importantly, the data continuity that is now being established (with the 2010 and 2012 waves and the proposed 2013 wave) would be lost, affecting the ability of researchers to draw inferences from trends in reported time use and well-being.

On the budget side, the marginal financial cost of adding the developed module to ATUS is relatively modest—about $178,000.[14] That said, it would be useful to perform a full accounting to assess the quality of survey results and any effects that the addition of the SWB module may have on the quality of the overall CPS and ATUS. At least in terms of respondent burden and response rates, these concerns would seem to be modest for the former and unfounded for the latter. Indeed, by design, the ATUS is asked of those who have rotated out of the CPS, and modules are asked after the core ATUS is completed. This design element prevents modules from impacting response to the core ATUS and CPS.[15] Because the SWB questions are the last thing the respondent hears, the impact on the core ATUS is expected to be minimal. Similarly, the SWB module cannot, by design, bias the core diary responses. On the respondent burden question, for the 2012 SWB module, average time spent was approximately 5 minutes, which adds up to an estimated 1,100 hours for the 12,800 respondents (*Federal Register*).

---

[14] The monetary cost of the 2012 module was higher ($273,000) as it included cognitive testing, data editing, interviewer training, and call monitoring activities by BLS.

[15] If ATUS interviewers indicated that the survey will take 5 minutes longer, addition of the module could affect people's willingness to participate (unit response rates). ATUS response rates have ranged from 52.5 to 57.8 percent. The response rates for 2010 (the first year of the SWB Module) was 56.9 percent.

3.3 VALUE OF A THIRD WAVE

A third wave of data collection will add significant information beyond what has been collected so far. Most obviously, another year for the survey means an increased capacity for researchers to enlarge samples by pooling data across years. For some purposes—for example, to look at well-being effects associated with changes in employment during recessions (only a small percentage of the population is unemployed) or to investigate differences across population subgroups—the number of observations needed to make valid statistical inferences well exceeds the annual sample size. This is especially true for comparing self-reported well-being score across smaller population subgroups. Almost all of the research to date using ATUS—which covers a wide range of topics, from household production, to work and leisure patterns, to childcare issues—has pooled data across years to increase the robustness of the statistical estimates.[16] The need to enlarge samples (pool data) will be true for research applications that rely on the SWB module of the ATUS as well.

Crucially, the 2012 module (the second wave) is only the first version of the survey that asks the overall life satisfaction (evaluative) well-being questions. In order to begin looking at sensitivity of measures and changes over time in these questions, at least one additional round of the survey—and ideally several more—are needed. A 2013 module would effectively double the sample size of respondents who have answered the evaluative well-being questions.

Fielding another round of the SWB module will also add to the accumulating evidence needed to determine the value of incorporating it into the ATUS (and possibly elsewhere) on something more than an experimental basis. More generally, continuing the module will encourage discussion of how measures of subjective well-being can play a useful role in assessing the effects of public policies. On the research side, a third wave of data may shed light on unanswered questions about survey issues, data quality, and reliability (e.g., nonresponse bias, question ordering, context effects). Other technical issues that could be studied include mode of administration effects (is reported well-being lower in face-to-face interviews than for telephone or Internet modes?); activation/valence (are positive and negative affect two ends of the same bipolar dimension or are they separable unipolar dimensions?), scaling (do populations from difference cultures or age groups systematically respond differently?), and memory bias (e.g., are negative events reported more or less frequently than positive events?).

---

[16] A bibliography of research that has used ATUS data can be found at http://ideas.repec.org/k/atusbib.html (accessed August 7, 2012).

A third wave of the survey could also be used to explore opportunities for experimentation designed to move toward an optimal survey structure, should the module become a permanent biannual ATUS supplement. Although it is unlikely that major changes could be made for a 2013 module, in the longer term it is certainly worth considering whether modifications could be made to increase its value. Examples of possible modifications to consider include

- Split sample surveys—one-half the respondents could receive one question while the other half gets another; this would be useful for testing such things as sensitivity to different scales and question wording.[17]
- Finding the optimal number of activities to ask about. It is not obvious that three activities is the optimal number of activities to include on the module. It may be useful to ask about hedonic well-being associated with more activities in order to increase the reliability of daily estimates. Importantly, sampling more episodes increases the power to examine activity-specific effects, which may be particularly valuable for addressing policy questions. Doubling or even tripling the number of episodes may be cost-effective, although that benefit would have to be weighed against considerations of participant burden and the potential impact on response rates.
- Selecting the "right" positive and negative emotion adjectives for module questions. Research supports the separation of positive and negative states but, more generally, should the module be focused more on suffering or happiness. The module could experiment with different adjectives and how interpretation varies across populations.
- Expanding coverage to pain and other sensations. There are no good conceptual criteria for differentiating between sensations and "pure" emotional states or for how the two link together. Intuitively, sensations are principally physiological states, in contrast to such feelings as anxiety, stress, and joy, which are principally subjective states.
- Additional or replacement questions for consideration. A possible example is adding a question or two about sleep, such as: "How many hours of sleep do you usually get during the week?" or "How many hours of sleep do you usually get on weekends?" The

---

[17] In its well-being survey, the UK's Office for National Statistics has used, or plans to use, split trials to test for such things as sensitivity to different scales, question wording, and order and placement of questions.

objective of such questions would be to find out if respondents' reports about behaviors/emotions—feeling happy, tired, stressed, sad, pain—are influenced by (chronic) sleep deprivation or other sleep patterns.[18] A methodological question is how well do people recall the previous night's sleep?
- Selecting among competing evaluative measures. Is the current Cantril approach, which is perhaps the most remote from affect measures, optimal? Alternative versions of the evaluative measure are common in the literature.

It would also be interesting to make modifications to the SWB module so that day-of-week effects could be tested for different domains—health, education, transportation, etc.

The merits of retaining some fraction of the sample for experimental work should be strongly considered, presumably not for 2013 but for subsequent years. One such experiment would be to determine sample sizes needed for subgroup analyses (e.g., day reconstruction method questions, which rely on some recall, are systematically answered differently by older and younger populations; in an aging society, it is important to be cognizant of these effects).

The ATUS SWB questions could be the model for a standard set of questions that could be added to other surveys. With effective data linking, this could yield a rich set of findings about the relation to SWB of a wide range of covariates. If such a strategy were adopted, the experience of the ATUS SWB module will provide insights about how questions might perform on health, economic, and other kinds of surveys; and for determining candidate surveys such as the National Health Interview Survey and the National Health and Nutrition Examination Survey, administered by the National Center for Health Statistics, and the Survey of Income and Program Participation, administered by the U.S. Census Bureau for adding modules. As noted above, there are potentially major advantages in having similar questions embedded across multiple surveys, especially as linking of microdata (including administrative) records becomes increasingly feasible.

In light of changing budgets and priorities and emerging alternative data sources (e.g., private label, digital, Web-based), the nation's statistical agencies have already begun to reexamine the content, modes, and structure of their surveys and data programs more intensively than ever before.

---

[18] This idea was raised by Mathias Basner, of the University of Pennsylvania School of Medicine, who noted that self-assessments of habitual sleep time overestimate physiological sleep time and that estimates of habitual sleep time based on ATUS overestimate self-assessments of habitual sleep times found in other population studies. Therefore, he suggested that it would be "very elucidating" to compare self-assessments of sleep time for the two questions above against estimates based on ATUS responses for the day before the interview day.

New scrutiny of what trends in society are important to measure (such as those recommended by the Commission on the Measurement of Economic Performance and Social Progress; Stiglitz, Sen, and Fitoussi, 2009) may give rise to new opportunities to refocus statistical program coverage (and the surveys on which they are built) and to move into new research areas surrounding SWB. Smaller-scale studies and data collections, such as the ATUS SWB module, are needed to help judge the value and feasibility of embarking on production of national-level SWB statistics, such as those under development in the United Kingdom. Moreover, determination of the place of measures of subjective well-being in monitoring the economy and society cannot be done without the data. The question of whether self-reported measures of well-being should one day be reported alongside more standard economic statistics, such as those for income and employment and for financial markets, is as yet unanswered.

A careful assessment of the data emerging from ATUS and the SWB module may help avoid mistakes if self-reported well-being statistics are ever produced on a larger scale. To the extent that evidence can be accumulated on the research and policy value of such data, a better basis for making these data collection and statistical program decisions can be established. The fact that the United States has a decentralized statistical system makes coordinating of the survey content related to subjective well-being a greater challenge than in countries with centralized statistics systems. However, it also affords the option of targeting development in the areas that are identified as the most relevant for policy and measurement—such as health, employment, or education—for which the argument is strongest for adding this kind of content. In light of these arguments, it is the view of the panel that the cost of the proposed 2013 SWB module is quite modest given its potential to inform decisions about potentially much larger statistical system investments.

References

Aguiar, M., and Hurst, E. (2007). Measuring trends in leisure: The allocation of time over five decades. *The Quarterly Journal of Economics, 122*(3), 969–1006, 1008.

American Time Use Survey User's Guide. (2013). *Understanding ATUS 2003-2011*. Report by the Bureau of Labor Statistics: Washington, DC. Available: http://www.bls.gov/tus/atususersguide.pdf [September 2012].

Archer, M., Paleti, R., Konduri, K., and Pendyala, R. (2012). *Modeling the Connection Between Activity-Travel Patterns and Subjective Well-Being*. Paper presented at the 92nd Annual Meeting of the Transportation Research Board, Washington, DC.

Boeham, J.K., and Kubzansky, L.D. (2012). The Heart's Content: The Association between Positive Psychological Well-Being and Cardiovascular Health." Psychological Bulletin, online April 17, 2012. Available: http://www.rwjf.org/pioneer/product.jsp?id=73919 (accessed September 7, 2012).

Christodoulou, C., Schneider, S., and Stone, A. (2012). *Validation of a Brief Yesterday Measure of Hedonic Well-Being and Daily Activities: Comparison with the Day Reconstruction Method*. Working Paper, June 4.

Deaton, A.S. (2011). *The Financial Crisis and the Well-Being of Americans*. NBER Working Papers 17128. National Bureau of Economic Research, Inc. Available: http://www.nber.org/papers/w17128 (accessed July 29, 2012).

Diener, E. (2006). Guidelines for national indicators of subjective well-being and ill-being. *Applied Research in Quality of Life, 1*(2), 151–157.

Diener, E., Kahneman, D., Tov, W., and Arora, R. (2009). Income's Differential Influence on Judgments of Life Versus Affective Well-being. *Assessing Well-being*. Oxford, UK: Springer.

Huppert, F.A. (2009). Psychological well-being: Evidence regarding its causes and consequences. *Applied Psychology: Health and Well-Being, 1*, 137–164.

Kahneman, D., and Deaton, A. (2010, August). High Income Improves Evaluation of Life but not Emotional Well-Being. *Proceedings of the National Academy of Science*.

Krueger, A.B., and Mueller, A. (2012). Time use, emotional well-being and unemployment: Evidence from longitudinal data. *American Economic Review, 102*(3), 594–599.

Krueger, A.B., and Stone, A.A. (2008). Assessment of pain: A community-based diary survey in the USA. Lancet, *371*(May 3), 1519–1525.

Meyer, B.D., and Sullivan, J.X.. (2009). Economic Well-Being and Time Use. Working paper, June 22.

Steg, L., and Gifford, R. (2005). Sustainable transportation and quality of life. *Journal of Transport Geography, 13*(1), 59–69.

Stiglitz, J., Sen, A., and Fitoussi, J.P. (2009). *Report by the Commission on the Measurement of Economic Performance and Social Progress.* Available: http://www.stiglitz-sen-fitoussi.fr/documents/rapport_anglais.pdf (accessed August 2, 2012).

Appendix C

Biographical Sketches of Panel Members

ARTHUR A. STONE (*Chair*) is distinguished professor of psychiatry and psychology and director of the Applied Behavioral Medicine Research Institute, all at Stony Brook University. He is also a senior scientist at the Gallup Organization, working with Gallup's well-being surveys. He specializes in the field of behavioral medicine, focusing on stress, coping, physical illness, and self-report processes and measures. He has been an executive council member for the American Psychosomatic Society, a research committee member for the American Psychological Association, and a past president and executive council member of the Academy of Behavioral Medicine Research. Dr. Stone serves on several national and international scientific advisory boards of survey studies monitoring the health and well-being of populations. He is a fellow of the American Psychological Association, the Society for Behavioral Medicine, and Academy of Behavioral Medicine Research, among others. He holds a B.A. degree from Hamilton College and a Ph.D. in clinical psychology from Stony Brook University.

NORMAN M. BRADBURN is the Tiffany and Margaret Blake distinguished service professor emeritus at the University of Chicago, where he also served on the faculties of the Department of Psychology, the Irving B. Harris Graduate School of Public Policy Studies, the Booth School of Business, and the College. He is a senior fellow at the university's National Opinion Research Center. Dr. Bradburn previously served as assistant director for social, behavioral, and economic sciences at the National Science Foundation. His research focuses on psychological well-being and the assessment of quality of life using large-scale sample surveys. He is a past president of the American Association of Public Opinion Research. He has an M.A. degree in clinical psychology and a Ph.D. in social psychology, both from Harvard University.

LAURA L. CARSTENSEN is professor of psychology, Fairleigh S. Dickinson Jr. professor in public policy, and founding director of the Stanford Center on Longevity, all at Stanford University. Much of her work has focused on socioemotional selectivity theory—a life-span theory of motivation. Her most current empirical research focuses on ways in which motivational changes influence cognitive processing. She is a fellow of the Association for Psychological Science, the American Psychological Association, and the Gerontological Society of America, and she is a member of the MacArthur Network on Aging Societies. She has received the Richard Kalish Award for innovative research, the Distinguished Career Award from the Gerontological Society of America, Stanford University's dean's award for distinguished teaching, and a MERIT (Method to Extend Research in Time) award from the National Institute on Aging. She has a B.S. degree in psychology from the University of Rochester and both an M.A. in developmental psychology and a Ph.D. in clinical psychology from West Virginia University.

EDWARD F. DIENER is the Joseph R. Smiley distinguished professor of psychology in the Department of Psychology at the University of Illinois at Urbana-Champaign and a senior scientist at the Gallup Organization. His research focuses on the measurement of well-being, temperament, and personality influences on well-being, as well as on theories of well-being, income and well-being, and cultural influences on well-being. He has served as president of the International Society of Quality of Life Studies, the Society of Personality and Social Psychology, and the International Positive Psychology Association. Among his many awards are an honorary doctorate from the University of Berlin and a distinguished scientist award from the International Society of Quality of Life Studies. Dr. Diener won the distinguished researcher award from the International Society of Quality of Life Studies, the first Gallup academic leadership award, and the Jack Block award for personality psychology. He received the American Psychological Association's distinguished scientist award in 2012 and the Association for Psychological Science's William James award for lifetime scientific achievement in 2013. He has a B.A. degree in psychology from the California State University of Fresno and a Ph.D. in psychology from the University of Washington.

PAUL H. DOLAN is professor of behavioral science in the Department of Social Policy at the London School of Economics and Political Science. He is also chief academic adviser on economic appraisal for the Government

Economic Service in the United Kingdom. Previously, he held academic posts at the universities of York, Newcastle, Sheffield, and Imperial, and he has been a visiting scholar at Princeton University. His research interests focus primarily on developing measures of subjective well-being that can be used in policy, particularly in the valuation of nonmarket goods, and in extending the ways in which the lessons from behavioral economics can be used to understand and change individual behavior. Dr. Dolan is a recipient of the Philip Leverhulme Prize in economics—awarded by the Philip Leverhulme Trust in the United Kingdom—for his contribution to health economics. He has served on many expert panels for various government departments in the United Kingdom. He has M.Sc. and D.Phil. degrees in economics from York University.

CAROL L. GRAHAM is Leo Pasvolsky senior fellow at The Brookings Institution, College Park professor in the School of Public Policy at the University of Maryland, and research fellow at the Institute for the Study of Labor (IZA) in Bonn, Germany. From 2002 to 2004, she served as a vice president at Brookings. She has also served as special advisor to the vice president of the Inter-American Development Bank, as a visiting fellow in the Office of the Chief Economist of the World Bank, and as a consultant to the International Monetary Fund and the Harvard Institute for International Development. Her most recent books are *The Pursuit of Happiness: Toward an Economy of Well-Being* (Brookings, 2011) and *Happiness Around the World: The Paradox of Happy Peasants and Miserable Millionaires* (Oxford University Press, 2010). Dr. Graham has published articles in a range of peer-reviewed journals, and her work has been reviewed in *Science*, *The New Yorker*, and *The New York Times*, among others. She is an associate editor at the *Journal of Economic Behavior and Organization*, among other journals. Her research has received support from the MacArthur, Tinker, and Hewlett Foundations and from the National Endowment for the Arts. She has an A.B. from Princeton University, an M.A. from Johns Hopkins University, and a D.Phil. from Oxford University.

V. JOSEPH HOTZ is the arts and sciences professor of economics in the Department of Economics at Duke University, research affiliate at the Institute for Research on Poverty at the University of Wisconsin–Madison, research fellow at the Institute for the Study of Labor, and research associate at the National Bureau of Economic Research. He also serves as a research affiliate at the National Poverty Center, the Gerald R. Ford School of Public Policy, and the University of Michigan. Previously, Dr. Hotz served as visiting scholar at the Cowles Foundation, Yale University, and at the Russell Sage Foundation. He was professor and chair of the Department of Economics at the University of California, Los Angeles. His areas of specialization include labor economics, population economics, and applied econometrics. He has a B.A. from the University of Notre Dame and M.S. and Ph.D. degrees in economics from the University of Wisconsin–Madison.

DANIEL KAHNEMAN is professor of psychology and public affairs, emeritus, and senior scholar at the Woodrow Wilson School at Princeton University. He is also the Eugene Higgins professor of psychology (emeritus) at Princeton University and a fellow at the Center for Rationality at The Hebrew University. Previously, he held positions as professor of psychology at the University of California, Berkeley, associate fellow at the Canadian Institute for Advanced Research, and visiting scholar at the Russell Sage Foundation. He is a member of the National Academy of Sciences, the American Academy of Arts and Sciences, the American Philosophical Society, and the Econometrical Society, and he is a fellow of the American Psychological Association. Dr. Kahneman is a recipient of the 2002 Nobel Prize in economics, as well as the distinguished scientific contribution award of the American Psychological Association, the Warren Medal of the Society of Experimental Psychologists, and the Hilgard Award for career contributions to general psychology from the American Psychological Association. He has a B.A. degree in psychology and mathematics from The Hebrew University and a Ph.D. in psychology from the University of California, Berkeley.

ARIE KAPTEYN is professor of economics and founding director of the Dornsife Center for Economic and Social Research at the University of Southern California. Previously he was a senior economist at RAND Corporation and director of its labor and population division. Before joining RAND, he held positions at Tilburg University in The Netherlands, including dean of the Faculty of Economics and Business Administration and founder and director of CentER, a research institute and graduate school. Dr. Kapteyn has held visiting positions at Princeton University, the California Institute of Technology, Australian National University, the University of Canterbury (New Zealand), the University of Bristol, and the University of Southern California. His research expertise covers microeconomics, public finance, and econometrics. He is a fellow of the Econometric Society, a member of the Netherlands Royal Academy of Arts and Sciences, and past president of the European Society for Population Economics. He has a B.A. and an M.A. in agricultural economics from State Agricultural University, Wageningen, an M.A. in econometrics from Erasmus University, Rotterdam, and a Ph.D. in economics from Leyden University, all in The Netherlands.

AMANDA SACKER is director of the ESRC International Centre for Lifecourse Studies in Society and Health and professor of lifecourse studies in the Research Department of Epidemiology and Public Health at University College London. Earlier she was research professor in quantitative social science at the Institute for Social and Economic Research at the University of Essex, England, and before that, principal research fellow at University College London. She holds numerous positions, including honorary research associate at the Institute for Social and Economic Research, member of the executive committee of the Society for Longitudinal and Life Course Studies, and member of the international journal *Longitudinal and Life Course Studies*. Dr. Sacker's research interests focus on life course epidemiology and inequalities in physical and mental health, with particular interest in the use of mixture models that combine categorical and continuous latent variable modeling techniques in longitudinal studies. She has a B.Sc. degree in psychology and a Ph.D. in psychology and statistics.

NORBERT SCHWARZ is provost professor of psychology and marketing at the University of Southern California, Los Angeles. Previously he was the Charles Horton Cooley collegiate professor of psychology at the University of Michigan, professor of business at the Stephen M. Ross School of Business, and research professor at the Institute for Social Research. Before that, he taught psychology at the University of Heidelberg, Germany, and served as scientific director of ZUMA, an interdisciplinary social science research center. His research interests focus on human judgment and cognition, including the interplay of feeling and thinking, the socially situated and embodied nature of cognition, and the implications of basic cognitive and communicative processes for public opinion, consumer behavior, and social science research. Dr. Schwarz is an elected member of the American Academy of Arts and Sciences and the German National Academy of Science Leopoldina. He has received the Heinz Maier-Leibnitz Prize of the German Department of Science and Education and the Wilhelm Wundt Medal of the German Psychological Association. He has a Ph.D. in sociology and psychology from the University of Mannheim and a Habilitation degree in psychology from the University of Heidelberg.

JUSTIN WOLFERS is professor of economics and public policy at the University of Michigan and a senior fellow at The Brookings Institution. Prior to these positions, he was visiting associate professor in the Department of Economics at Princeton University, associate professor of business and public policy at the Wharton School of the University of Pennsylvania, and assistant professor of political economy at Stanford University. He holds numerous other positions including research associate at the National Bureau for Economic Research and senior scientist at the Gallup Organization. Dr. Wolfers' research interests include law and economics, labor economics, social policy, political economy, macroeconomics, and behavioral economics. He is the recipient of numerous awards, including the Wharton M.B.A. core teaching award and the excellence award in global economic research from the Kiel Institute, Germany. He is also a columnist for *Bloomberg View* and a regular commentator on American Public Media's Marketplace radio program. He has a B.A. in economics from the University of Sydney and A.M. and Ph.D. degrees in economics from Harvard University.

COMMITTEE ON NATIONAL STATISTICS

The Committee on National Statistics was established in 1972 at the National Academies to improve the statistical methods and information on which public policy decisions are based. The committee carries out studies, workshops, and other activities to foster better measures and fuller understanding of the economy, the environment, public health, crime, education, immigration, poverty, welfare, and other public policy issues. It also evaluates ongoing statistical programs and tracks the statistical policy and coordinating activities of the federal government, serving a unique role at the intersection of statistics and public policy. The committee's work is supported by a consortium of federal agencies through a National Science Foundation grant.